Technology

,d.

TOTAL AREA
NETWORKING

About the Wiley — BT Series

The titles in the Wiley-BT Series are designed to provide clear, practical analysis of voice, image and data transmission technologies and systems, for telecommunications engineers working in the industry. New and forthcoming works in the series also cover software systems, solutions, engineering and design

---

Other titles in the Wiley-BT Series:

TOTAL AREA NETWORKING
ATM, IP, FRAME RELAY AND SMDS EXPLAINED

Second Edition

John Atkins

Mark Norris
BT, UK

JOHN WILEY & SONS
Chichester • New York • Weinheim • Brisbane • Singapore • Toronto

National   01243 779777
International (+ 44) 1243 779777
e-mail (for orders and customer service enquiries): cs-books@ wiley.co.uk
Visit our Home Page on http://www.wiley.co.uk
                or http://www.wiley.com

Other Wiley Editorial Offices

John Wiley & Sons, Inc., 605 Third Avenue,
New York, NY 10158-0012, USA

WILEY-VCH GmbH, Pappelallee 3,
D-69469 Weinheim, Germany

Jacaranda Wiley Ltd, 33 Park Road, Milton,
Queensland 4064, Australia

John Wiley & Sons (Canada) Ltd, 22 Worcester Road,
Rexdale, Ontario M9W 1L1, Canada

John Wiley & Sons (Asia) Pte Ltd, 2 Clementi Loop #02-01,
Jin Xing Distripark, Singapore 129809

British Library Cataloguing in Publication Data

A catalogue record for this book is available from the British Library

ISBN 0-471-98464-7
Typeset in 10/12pt Palatino from the author's disks by Vision Typesetting, Manchester
Printed and bound in Great Britain by Bookcraft (Bath) Ltd
This book is printed on acid-free paper responsibly manufactured from sustainable forestry,
in which at least two trees are planted for each one used for paper production.

Contents

Preface

In a world more and more driven by technology, why would anyone in their right mind want to sacrifice good drinking time to write a book? We have asked ourselves this question many times, and we think we have an answer. Here it is.

There are sales people and market makers, who have little time away from the cut and thrust of the sharp end to ingest volumes of technical data, and there are armies of clever and inventive technocrats who delight in crafting yet more intricate and complex technology. The two have little time to communicate. Even if they did, it is unlikely that they would speak the same language. The aim of this book is to bridge that gap.

We have tried to provide a readable and accessible account of a key area of technology. One that affects us all and one that will have a major impact on our ability to compete, as individuals, as organisations and as nations. With the Information Age upon us we have distilled the basics. We think there is enough of a technical grounding here to enable you to discriminate technobabble from fact, and enough analysis to put deep knowledge into context.

Our approach can be summarised in two phrases: 'How do you know?' and 'So what?'. We have scoured volumes of standards, product specifications and the like to distil our facts; that is how we know. The 'So what?' you can judge for yourselves. We have explored our vision of the Information Age with many people; we think it is a good guess.

So, technocrats, see how you can best apply your wealth of knowledge in a way that users appreciate. Users, debate with your technocrats to get what you want, based on a clear understanding of what the technology can actually do for you. We would both drink to that!

Mark Norris
and
John Atkins

Foreword

As a journalist, I know the importance of networks: without colleagues, friends, associates and a thick book of contacts, I simply could not work. For at the heart of good networking is quality information. However, as I put together the series 'The Network' for Radio 4, I realised that the acquisition, the manipulation, the analysis and the dissemination of information is vital not just for people who work at the sharp end of the media, but for everyone. Networks of one kind or another are vital for almost every human activity: whether to get cash out of a hole in the wall, or to decide what to buy with the money, based on discussions with friends and reading the latest consumer guides.

Networks, nowadays, also need to form larger networks, an infuriating habit when trying to create neatly segmented radio programmes, and as a consequence it's often hard to determine where one network stops and another one starts. The quality information you or I might be looking for in a weather forecast, say, is whether we should pick up a brolly on the way out, or not. This is a decision derived from a network of satellites, of land and sea based weather stations, of supercomputers, a national and local network of broadcasters, and possibly most importantly a quick chat on the way out with a loved one who happened to catch the most recent forecast! A whole web of interconnections, a pile of protocols and everything in bandwidth from a chat to television channels.

Yet weather forecasting highlights a crucial element of powerful networks: to the person who needs the information the network has to be transparent. The telephone system is the most complex global machine that we humans have ever built, and for the vast majority of us we just pick up a 'phone and dial—anywhere. The incredible complexity of the international telephone system has been rendered completely invisible, a feat which seemed always to have eluded computer networks. However, this could all be about to change. The reticence with which computers speak to each other is being eroded. The inflexibility of computer networks, providing only what they could and not what we needed, has had its day. In this book Mark and John have pulled together the technologies that are starting to allow the networks of the future to fulfil their past promises: teleworking, remote consultancy, video conferencing, home shopping and banking. Technologies that will camouflage the complexity of global computer networks, allowing us to be just a keystroke away, an icon and a click nearer to the information we will all need to function in the next century.

Peter Croasdale
Editor, *'The Network'* and *'Science Now'*
BBC Radio

About the Authors

Mark Norris has 20 years' experience in software development, computer networks and telecommunications systems. Over this time he has managed dozens of projects to completion from the small to the multi-million pound, multi-site. He has worked for periods in Australia and Japan and currently manages BT's broadband support systems development. He has published widely over the last ten years with a number of books on software engineering, computing, technology management, telecommunications and network technologies. He lectures on networks and computing issues, has contributed to references such as Encarta, is a visiting professor at the University of Ulster and is a fellow of the IEE. Mark plays a mean game of squash but tends not to mix this with Total Area Networking.

John Atkins is now a freelance consultant, having previously been a senior engineering advisor with BT. He has over 30 years' experience in telecommunications networks, which spans analogue and digital switching systems for both voice and data services, exploiting both circuit and packet modes of operation. His current interests lie mainly in the technology and standards for networking and internetworking, especially how they can be used to exploit the new opportunities arising from increased liberalisation in telecommunications. When he is not writing or consulting John plays the blues on his positively edible Gibson 335.

Acknowledgements

The authors would like to thank a number of people whose help and co-operation have been invaluable. To those kind individuals who volunteered to review early drafts: Nick Cooper, Ian Gallagher, Barry Hinton, Mark Jakeman, Darius Karkaria, Peter Key, Trevor Matthews, Paul Mylotte, David Packer, Malcolm Payne, Lucy Powell, Steve Pretty, Dave Quinn, Bonnie Ralph, Sinclair Stockman, Dave Sutherland, Alan Topping and Steve West. Their observations, guidance and constructive criticism were always valuable (if, on occasions, painful) and have done much to add authority and balance to the final product.

Thanks are also due to our many friends and colleagues in the telecommunication and computing industries, whose experience, advice and inside knowledge have been invaluable. Finally, a special thank you to Chris Bilton for his wisdom, incisive comments and thorough reviewing.

1

The Information Age

On the information superhighway there will be few speed limits and no turning back

Financial Times, 1994

Information is vital to modern business. Effective use of information is, at least, an issue of money; more often it is one of survival, and survival is not compulsory.

The way in which information is generated, stored, transmitted and processed is changing in a way that will revolutionise many people's work practices in the not too distant future. The end of the 20th century will be recognised as the dawn of the Information Age, as significant a change as the earlier Industrial and Agricultural Revolutions.

The user end of this revolution is already familiar to most of us. As before, it has been driven by the emergence of powerful new technology—explosive growth in personal computers and the increasing use of local networks has given many the freedom to do a great variety of tasks without moving from a single location (Cook *et al* 1993).

Further to this, expectations of having the world at your fingertips have been raised. A global data communications market growing at 30% per annum is providing the necessary infrastructure. On top of this, ideas of 'teleworking', 'virtual teams' and indeed 'virtual corporations' have been shown to be viable through a combination of computing and telecommunications (Davies, Sandbanks and Rudge 1993).

In practice, though, the information age is not as cut and dried (nor as straightforward) as portrayed by some prophets (and many suppliers). Behind the user's ideal of 'a world at your fingertips' lies a complex infrastructure. The ways in which this can be assembled and controlled must be understood so that appropriate options and configurations are used. Central to this are the new data communications technologies—Frame Relay,

The Information Age will see . . .

Total Area Networking
the deployment of high-speed data communications to allow currently localised facilities to become as distributed as the telephone network. It provides the information transport to enable superconnectivity

and

Superconnectivity
allowing the user to see a distributed set of network-based resources as one. It is the embedded intelligence that takes the distance out of information. It enables electronic trading.

Figure 1.1   Key concepts for the Information Age

SMDS and ATM. These, along with the established Internet, provide the basic enablers of the information age: together, they hold the key to superconnectivity (see Figure 1.1). Yet these essential technologies are often portrayed as mutually exclusive rather than as options to be combined, as the situation demands, to meet a business requirement.

A focus on the technical detail of specific solutions, rather than on how they combine to good effect, is not altogether surprising at this early stage of the information age. The main aim of this book is to show how the new data communication technologies contribute to the user ideal outlined above; in particular, how they are likely to be used to extend the facilities now available on local networks across the globe.

The central message is that the differentiation between local area (often private) networks and wide area (usually public) networks will gradually disappear. What the user *actually* wants is total area networking and this will be the enabler for the Information Age.

This may all seem to be no more than an escalation of what already exists; but there are significant implications to **Total Area Networking**, some driven by user needs, others driven by the technology itself.

The first implication is derived from looking at the innovation that has been evident over the last ten years through the use of local area networks. With a shift from local to total area networking, there is likely to be a matching shift from local to global business operations. This trend is already well established (Naisbitt and Aburdene 1986) and will be fired by relevant technology.

A second important implication is that the removal of barriers, imposed at present by disjoint networks, will make it viable to treat the network as a resource. Once this has been accepted, users will demand ever greater data communication capacity as a basis for their business. This will enable more distributed working. And this, in turn, will fuel the demand for greater connectivity of remote systems. The business norms will rapidly shift from *'write a cheque'* to *'use BACS'*, from *'post a letter'* to *'send email'*.

A little understanding of the key issues that need to be addressed in practical implementation will ease the migration to total area networks. We will spend the next part of this chapter explaining the context in which the total area network is evolving. Following this, we will briefly look at the implications of the total area network, specifically those for the near future.

1.1   THE IMPACT OF THE INFORMATION AGE

Over the last 25 years or so a number of factors have changed the nature of business competition. The particular drivers have varied from one industry to the next, but there are common themes. Among these are increased similarity in available infrastructure, distribution channels and business practice among countries. These, along with a fluid capital market that allows large-scale flows of funds between countries, have led to a situation where information is a vital resource and its global availability is a prerequisite to competitiveness. The point is readily illustrated. Not many years ago, aeroplanes were the major cost in the airline business. These days, they are easier to replace than the associated flight booking systems: information, shared over a wide area, is now the key resource.

The capability and availability of reliable telecommunications is, to a large extent, the enabler for organisations to generate and share information effectively, irrespective of location. It is now possible to transmit volumes of detailed and structured information across the globe in seconds, with little or no corruption. The information industry is here, and it is now widely predicted that co-ordination among a network of activities (e.g. design, marketing, production, etc.) dispersed world-wide is becoming a prime source of competitive advantage—the business battleground of the second millennium (Porter 1986).

As the information age matures it will change the way we work, both as individuals and as organisations (Ohmae 1992). For the information-intensive company there will be a greater choice of the fastest, cheapest, easiest way to deliver the goods by:

- being flexible in responding to competition by sharing information between regions

- reducing costs through the selection of the cheapest resource, wherever located

- enhancing effectiveness through the selection of the best set of resources, irrespective of location.

The impact on the individual will be no less dramatic. There will be less attachment to specific locations as it becomes easier to establish virtual teams

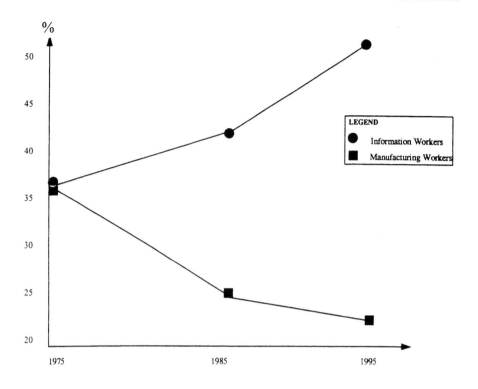

Figure 1.2 The move from manufacturing to information work

—groups of individuals who can communicate and share information as they work together unaffected by the tyranny of distance. The rise of the 'information worker' has been evident for some years now, as illustrated in Figure 1.2.

This trend is likely to continue and, with total area networking, it will become increasingly difficult to discriminate where the boundaries between consumer, information (or knowledge) worker and business operations are. As shown in Figure 1.3, the future will be characterised by co-operation.

The operational impact of the information age and the 'new rules of the game' for virtual teamworking will be covered in later chapters.

1.2 VALUE SEEKERS AND ECONOMY SEEKERS

The Information Age will affect each business in a different way and at different times. For some, business survival depends on being in the vanguard of new technology. Others can quite safely carry on as they are with well established information and communication systems (Porter 1987).

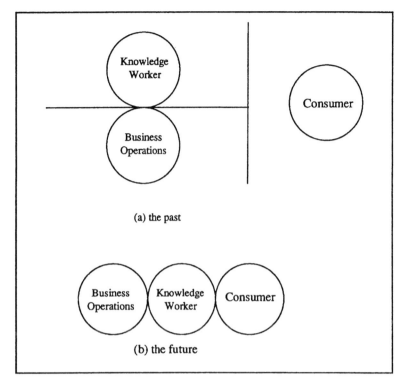

Figure 1.3   The network will remove boundaries

These two extremes can be categorised, respectively, as value seekers and economy seekers.

It is important to establish whether you are a value seeker or an economy seeker. This should dictate the level of interest (and, probably, expenditure) that you invest in the emerging technologies explained here. We can further characterise the two main categories as follows.

> Value seekers are looking, primarily, for functionality and they are prepared to pay for it. Typically, they are the larger companies, often multinationals with a global presence and, hence, global networking needs. Many of these companies operate their own private networks, for both voice and data services. The general trend among value seekers is to focus on core business and to sub-contract (outsource) their communications provision.

> Since they tend to be large, there are relatively few value seekers. Economies of scale are important to them to keep costs down and there is intense competition between suppliers to provide what they want. The trend, therefore, is towards early delivery, even when standards have yet to be established. Indeed, the value seeking sector has led to the creation of new fora for rapid standards implementation (e.g. Frame Relay Forum,

ATM Forum, Internet Engineering Task Force, Telecommunications (was Network) Management Forum). These groups, usually supplier-led, aim to agree early intercepts of standards still being developed. A key feature of this sector is, therefore, rapid rate of evolution. The name of the game is to pick the latest technology at the earliest opportunity to keep ahead of the competition.

By contrast, the economy seekers tend to have more modest requirements, at least in terms of technology. They are a much larger group, numerically if not in total size, and they tend to be much more cost-conscious. Economy effected through the use of commodity items is important, and this sector is usually served by well established communication offerings—public switched networks such as ISDN. Developments here await the relevant standards that allow network components to become commodity items (both to buy and to own): evolution is slower and more predictable.

It should be recognised that many companies will be both value seekers and economy seekers. For example, a large multinational bank will be a value seeker in terms of its requirements for the global interconnection of its management information systems. But when it comes to supporting its high street outlets, it will be an economy seeker.

From the user's point of view, it is most important to establish which category applies. and where (Huber 1984). Behaving like a value seeker when you do not need to can be costly. Worse still, the adoption of an economy-seeking approach in an area of the business that requires the leading edge of technology is likely to result in the rapid demise of that part of the business.

1.3   CONVERGENCE AND COLLISION

The delivery of total area networking relies on the convergence of two traditionally separate communities. One is the network providers, who look to higher bandwidths and embedded intelligence to add value to their already widely distributed resource. The other is the computing suppliers, who aim to distribute their processing applications ever wider.

Despite converging on the same goals, they have very different ways of achieving them. This is not surprising, bearing in mind their disparate backgrounds. Distributed computing and intelligent high-speed networks represent a single target separated by different languages, concepts and ways of operating.

The force for convergence will be the user, and for this to be effective it is important to appreciate the differences of approach. To that end we explain briefly the *modus operandi* and driving forces behind the two.

The earliest demand for switched data services, during the 1960s and early 1970s, arose largely from the high cost of computers. To make the most effective use of the expensive data-processing equipment, time-sharing by remote terminals was introduced with the ubiquitous telephone network

(PSTN) being pressed into service to provide the necessary switched access. As computer technology developed it quickly became clear that the PSTN would not be able to support many of the new applications and that dedicated switched data networks would be needed.

The 1970s and 1980s have seen the implementation of these switched data networks, both as private networks built as company infrastructure, and as public networks offering a wide range of services. Agreements on international standards promoted rapid implementation, and most countries are now served by public data networks with international interworking providing global coverage. It is important to realise that these wide area network (WAN) developments have been driven mainly by the 'telecommunications community', the public telecommunications operators (PTOs).

The same period has also seen the stranglehold of the mainframe broken by explosive growth in the number of personal computers (PCs). Today computing power is cheap and very widespread and data communications are motivated more by the need to share information than to share equipment. Without fast and reliable access to information many modern organisations would fold—in days! To service this need the 'computing community'—the suppliers and users of computing equipment—have developed local area networks (LANs) to give terminals access to the company's hosts, and increasingly to provide terminal–terminal connectivity for distributed computing applications.

LANs provide interconnection over comparatively short distances (a few hundred or few thousand metres, depending on type) and are basically designed to cover single-company single-site situations. Large organisations inevitably have many sites, often distributed globally, and one of their key requirements is interconnection of the LANs over wide area distances. At the time of writing, wide area interconnection of LANs to create virtual private networks and intranets is one of the fastest growing markets of all. With the development of new distributed applications, including multimedia, this trend is clearly set to continue.

The problem is that the telecommunications community and the computing community are worlds apart! They developed from different beginnings and have different cultures, drivers, dynamics and players. To understand the directions and roles of the new data network technologies it is important to see them in the cultural as well as technical context.

The telecommunications community, having grown from the telephone, has its roots in the distant past and for many years enjoyed the position of monopoly supplier. Currently the trend almost everywhere is towards liberalisation and (often regulated) competition, but in most countries free market conditions are not yet the norm. The tradition in telecommunications is for implementation to follow international agreement on appropriate standards. This is necessary to ensure international interworking of networks and services, but it generally means that new developments take place slowly.

The computing community, on the other hand, is comparatively young (the first commercial computers date from the late 1950s) and has generally escaped the straitjacket of monopoly and regulation. The main driver is market forces. Product differentiation is therefore a key ingredient for success, and proprietary system architectures and protocols have been developed by different suppliers in their search for competitive edge. This means that advances have been very rapid, but the lack of standardisation also means that interworking between different suppliers' products is problematic or even impossible. The responsive nature of the computing community has of course usually provided *ad hoc* interworking solutions—the price the user pays is usually one of increasing, and sometimes unmanageable, complexity. Interworking LANs over wide area networks further adds to complexity because the two communities have developed different and often conflicting protocols.

Escape from this complexity trap depends to a large extent on the convergence of the telecommunications and computing communities. The new data networking technologies offer a platform for this convergence in supporting data communications that span the local area and the wide area seamlessly, to provide what we have called Total Area Networking.

Given that two previously isolated worlds could well collide, rather than converge, and that some will be at the point of impact, others at the periphery, you will need a route map. The last part of this preparatory chapter aims to set things in context (to position the value seekers, the economy seekers and those in between) by outlining a framework of networking styles. A point of reference.

Before explaining this framework, which consists of seven distinct (but not mutually exclusive) networking styles, we should establish some scale for the route map. The main point here is that the majority of users will require little more than the first three or so styles to support the way they work. For them, the ISDN (Griffiths 1992) is the network of the future. It provides all the bandwidth and connectivity that they need. Even if they do migrate towards more distribution, they are likely do so only slowly. In the early days of the Information Age, there will be only a few value seekers.

1.4 THE MAGNIFICENT SEVEN

Seven styles of networking can be defined to give a basic spectrum that extends from isolation through to a total area network. The model—illustrated in Figure 1.4—is based on the notion that there are three main components that characterise the way in which information is handled. These are:

- presentation services (focused on the user, what they see)

- data services (focused on data storage and sharing, where resources are)

- application logic (provides interworking services, the physical connection and logical binding of the first two).

This is an oversimplification, but it has its uses. Figure 1.4 shows how the

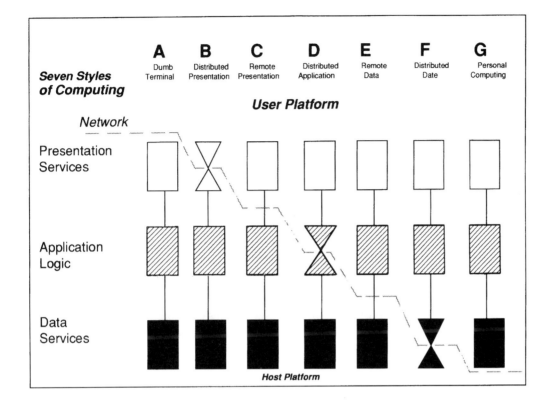

Figure 1.4 Seven basic styles of computing over networks

network can cut between components, or through a component, resulting in the seven possible divisions, each of which is explained below. This is an abstract representation: real systems may be split by the network in more than one place.

Above the network line, the functions reside locally; below it, they are remote. The distribution of functions between the two is what determines the requirements of the network (e.g. bandwidth, naming/addressing, etc.).

Personal computing

Here, all of the functionality and data is on the user platform. There is no communication requirement and hence no need for a network.

Dumb terminal

All logic runs on a central processor, with user access via 'dumb' terminals. In reality, the dumb terminals in use today are not completely dumb, as they provide a very simple presentation service (based on a general-purpose message set such as the ANSI standard, DEC VT100 or IBM 3270). In some cases the user end contains some application 'logic', in the sense that function keys can be downloaded from an application. An important parameter of this service is whether it sends keystrokes one at a time (character mode) or in blocks when the user hits a triggering key, such as ENTER. A permanent low-speed connection (e.g. a modem link) is required between terminal and host to support interaction.

Distributed application

There is application logic both locally and on remote processors. The classic example is where an application makes use of a remote procedure call (RPC) mechanism to invoke services as they are required from wherever they are available. Small volumes of high-speed traffic are required here, with a typical instance being a user with access to several servers connected on a local area network.

Remote presentation

The presentation service runs entirely on the user's terminal. This implies that remote applications do not control the manner of presentation of their data, which is rare. An accurate fit for this model might consist of an application that outputs pairs of numbers to a graphic server that draws graphs, histograms and so on. In practice, you might say that X-Windows applications fit into this category, or systems that have a very small amount of presentation-related application logic at the front-end. The links from user to remote services are small in terms of volume, but response times do have to be assured for the service to be acceptable.

Distributed presentation

The presentation service is provided partly by the local terminal and partly by the remote processor. An example of this is where the processor believes it is driving a dumb terminal, but this has been replaced by an intelligent machine at the front-end. The requirements on the network in this case are similar to those for the centralised application.

Remote data service

The application is either on the user's terminal or on a local server. The data is on remote processor(s). An example is an application of making query calls to a remote database management system (DBMS). In this instance there are potentially large volumes of data to be transmitted from a variety of sources. Fast response time and high peak bandwidth need to be assured to various remote locations.

Distributed data services

Data management is partly local and partly remote. This may be supported by facilities such as the network file system (NFS), which has Client and Server components to make a remote file system appear to be local to the application, hiding the network boundary. (Note that it is the service that is distributed, the data is remote.) As with the previous case, there is a requirement for a flexible high-speed network.

The details of the technology that can be used to populate and implement the above styles can be found in the appendixes, if required.

In general, as the sophistication of the service rises, so do the communication requirements. The last few of the above options require variable amounts of data between a number of locations, very much the 'value seeking' end of the spectrum. This could be achieved with a (potentially huge) number of permanent circuits, but both cost and lack of flexibility prohibit this in practice. It is this style of application that will drive total area networking, especially given the increasingly global scale of modern business.

1.5 PREPARING FOR SUPERCONNECTIVITY

So far we have outlined the central role that information networks are likely to play in the not-too-distant future and have laid out some basic concepts to help to describe where we are. How we move towards the Information Age is the subject matter of the remainder of the this book. Before explaining the structure of the technical detail, we dwell briefly on the context in which it needs to evolve (at least, from the user's point of view).

As indicated earlier in this chapter, we presently have three players whose futures lie in total area networking:

- the users, who need it to enable their businesses

- computer suppliers, who see an opportunity to sell more machines and applications

- the public network operators, who recognise escalating demand for flexible bandwidth.

This three-party arrangement is rather inelegant, in that, potentially, it leaves the user suffering from the confusion of having to reconcile confusing (at least) and conflicting (most likely) sales pitches. In effect, the user sits between two suppliers and has to resolve and integrate their offerings.

The central aim of this book is to establish a more controlled customer/ supplier arrangement in which the users can articulate their requirements in terms against which the suppliers can deliver. This does not mean that complexity can be ignored—it is fundamental to competition, is driven by differentiation and will always be there, especially for the value seekers. The intricacies and subtleties inherent in total area networking can, however, be controlled once an appropriate level of understanding has been established; and this places the user, not the supplier, in the driving seat.

The next chapter will explain how networks and the applications that they enable have evolved over the last fifty years. The fundamental issues that need to be addressed in moving from the piecemeal to the integrated and to the distributed will be illustrated. A more detailed picture of total area networking will be assembled in the context of how data communications have evolved.

Most of the subsequent chapters provide the technical detail to underpin the ideas developed in Chapters 1 and 2. Each is structured to present the key ideas up front, with layers of detail, if required, following on. This should allow you to establish the big picture quickly and to dive deeper later.

Chapter 3 will explain the detail of the first of the three emerging technologies, Frame Relay. The basic concepts and applications of Frame Relay will be covered to give a clear view of what it is, how it works, where it came from and why. The practical uses of Frame Relay will be explained and reference to its defining standards will be made.

Chapters 4 and 5 will cover similar ground for SMDS and ATM, respectively. Again, the technical aspects will be supplemented with an explanation of what they can be used for and how they fit into a wider networking picture.

Although focused on distinct technical subjects, each of Chapters 3, 4 and 5 aims to show their relative strengths, weaknesses and characteristics; in effect, why, when and how each should be used.

Chapters 6 and 7 illustrate application of the technology. The former takes a telecommunications perspective and describes how intelligent network principles can be used to build a Total Area Network. The latter is the computing view—an intranet.

Chapter 8 moves to the often forgotten area of network management. The concepts, requirements, guidelines and tools for keeping a complex network in tune with user needs will be covered in some detail. The realities of managing total area networks will be illustrated with reference to the some of the commercial tools that are now becoming available.

The final chapter will draw together the technical and operational threads to indicate what is possible now, what will be possible in the near future and how best to move between the two. The impact of total area networking on business and on the individual will be revisited.

Three appendixes provide much of the background required for the main text. Basic concepts in telecommunications and distributed computing are explained in Appendixes 1 and 2, respectively, as a grounding and/or reference for the main core of the book. The evolution of network services and elements from the operator's point of view is explained in Appendix 3.

1.6 SUMMARY

This chapter has looked at the forces that are driving the evolution of information networks. The main feature that is drawn out is that information will be *the* currency of future business.

An Information Revolution, as significant as the previous Industrial and Agricultural Revolutions, is inevitable. Trends towards global operations and reliance on distributed working are already evident, and are likely to characterise the way in which people work in the near future.

Given this context, there is a need for an infrastructure that matches these new requirements. The idea of total area networking is introduced here as the key to this. The concept is a simple one—that the distinction between local and wide area networks will disappear as data communications services, such as Frame Relay, SMDS and ATM are installed. And this will have a dramatic effect on the way in which information and networks are used and managed.

Many of the elements of Total Area Networking are already available but they need to be assembled as a whole, rather than as disparate 'point' solutions. A vital issue for the front-runners in the Information Revolution will be how to capitalise on new technology and how to articulate how it should be deployed to support their needs.

We have set the context and background for technical detail to follow. Since the aim of this book is to put a confusing raft of technical developments into an order relevant to the end user, the points made here should overlay all that follows.

REFERENCES

Cook, P. *et al.* (1993) *Towards Local Globalisation.* UCL Press.

Davies, D., Sandbanks, C. and Rudge, A. (1993) *Telecommunications after AD2000.* Chapman & Hall.

Gray, M., Hodson, N. and Gordon, G. (1993) *Teleworking Explained.* John Wiley & Sons.

Griffiths, J. (1992) *ISDN Explained.* John Wiley & Sons.

Huber, G. P. (1984) The nature and design of post industrial organisations. *Management Science,* **30.**

Naisbitt, J. and Aburdene, P. (1986) *Reinventing the Corporation.* Futura Books.

Ohmae, K. (1992) *The Borderless World.* Harper Collins.

Porter, M. E. (1986) *Competition in Global Industries.* Harvard Business School

Press.

Porter, M. E. (1987) From competitive advantage to corporate strategy. *Harvard Business Review*, May–June.

Strategic impact of broadband communications in insurance, publishing and healthcare. *IEEE Journal on Selected Areas in Communications*, **10**, December.

2

The total area network

The imperatives of technology and organisation, not the images of ideology, are what determine the shape of economic society

J. K. Galbraith

It was not very long ago that users of the UK postal system had a choice of postboxes, one marked 'local', the other 'national'. To use this system effectively, you had to know how it worked—that local letters would arrive more quickly, but would be delayed if misdirected through national routes.

The current picture of information networks is analogous. The onus is placed on the user to find out how the system works so that he or she can select optimal (or even, viable) routes and resources. But this situation is changing, and fast. Just as the postal service evolved to handle routing so that it was transparent to the user, so will information networks. In broad terms, the nature of the evolution will be similar—less reliance on user knowledge, more intelligence embedded in the network.

This chapter traces the evolution of information networks from early times, through to the current day and on to the near future. This puts some flesh on the bare bones of Chapter 1 and explains, in some detail, the trends and drivers that are shaping the emerging networks for the information age. We build up here a picture of what users are likely to require, and go on to give an overview of the new network technologies that promise to meet these requirements.

The key theme of this chapter is that many more people will become information-intensive workers over the next few years. Some already are, and a partial blueprint for the operating environment of the future already exists: there is a significant community of specialists who rely heavily on facilities such as the Internet and the World Wide Web to do their jobs. A faster and more robust infrastructure, provided through Total Area Networking and Superconnectivity, will speed popular adoption of information-intensive working and will, in turn, generate more innovation.

But there is more to the future than advances in technical capability. Part of this chapter considers the impact that the information revolution will have on the way in which people work. Again, there is a blueprint for this in that existing information workers have established ways of working that capitalise on remote access to information and facilities. With physical boundaries minimised and distribution as the norm, teams (and indeed organisations) can exist as virtual entities. Some of the 'new rules of the game' that will apply in the Information Age and to the virtual organisation are explored here.

To close this chapter we introduce some of the technical detail of the new datacommunication technologies that constitutes the technical core of the book.

2.1 THE STORY SO FAR

History shows that advances in communication have not been driven simply by new technology. Sure enough, this has proved a significant factor, but political, economic, social and regulatory issues have also played a major role. To really understand the current position and the likely future, we need to reflect on how we got where we are today, what is possible and what are the drivers for change are (Monk 1989). This section takes a few snapshots of global communication in the years since electronics began to provide an alternative to paper as the mass communication medium.

Perhaps the earliest relevant snapshot of information working (this is defined here as communications + processing, as opposed to telecommunications, which probably predates the Egyptians) would not come until the 1960s. Prior to this, networks had enabled information to be sent around the globe but the processing was predominantly in the hands of the person receiving the message. By the early 1960s many companies had realised that computers could quickly and accurately process large amounts of information.

Computers soon achieved the status of being a valuable business tool and were entrusted with significant amounts of important data. Also, in anticipation of things to come, they were being equipped with the means to communicate over public telephone lines. This made co-operative working between London, Los Angeles and Sydney viable as a routine part of business.

Data could be transported from one location to be used at the next. The process was neither fast nor elegant, typically consisting of sending source data from Sydney to LA, accepting a delay for processing at LA and awaiting a result to be sent on to London. Overall, the operation tended to be slow, costly and error-prone, a specialist exercise to be invoked only where necessary. Nonetheless, trust in computers as a support to business operations was established.

By 1980 the situation was altogether more reliable and speedy. Instead of shipping data over telephone lines, it was possible to interact directly with a distant computer using remote log-on facilities and dedicated data network links. Accessing data from a variety of sources was still slow and laborious, though, as multiple log-ons, file conversions and data transfers were usually required. Even so, experts could do a lot from the computers that were beginning to be placed on their desks.

Electronic mail was beginning to be an accepted and a well used means of communication. It added a human dimension, and the previously rather anonymous nature of information working started to give way to a more cooperative style.

In addition to this, public data services (such as the X.25 data network) provided reliable transmission links. Our global scenario would, by this time, probably be managed out of Sydney with the final result being emailed to London within a few hours, as opposed to a couple of days.

By the 1990s, business information had become the international currency of the global economic village. Its flows now dictate customer orders, product inventories, accounts payable, revenue and profit. The evolution in computing and telecommunications technology has enabled the rapid and reliable exchange of information on a global basis. International airline booking systems and automatic teller machines are an accepted part of everyday life.

The once complex operation described earlier is now a simple matter of London consulting a closed user group to which Sydney has posted the required information.

Simplicity of use has, however, a downside. It has been bought at the cost of huge complexity behind the scenes: the user's view of 'a world of information on one screen' is the result of a complex set of co-operating elements. The very fact that computing and telecommunications both play major roles in this means that the provider of an information network has to understand a diverse range of components, in particular how they can be assembled and configured to meet a particular set of needs. In practice this usually entails a mix of private and leased facilities. The former would be processors, databases, local networks and the like; the latter public network services, subscription data services, etc.

A typical network would consist of one personal computer (PC) per user on a set of linked local area networks (LANs). Each PC would be equipped with a set of software packages to allow users to access a wide range of facilities (printing, private and shared files, common applications, mail, etc.) hosted on different machines, some local, some remote (Karimi and Konsynski 1991).

The processing power available to the individual has now reached the level at which most information-based operations can be carried out without moving from the desk. Information can be found, collated and used as easily from a source on the other side of the world as in the next room, and this flexibility has come to drive the types of organisation that work with (and increasingly rely on) information as a resource.

The current situation is very much characterised by the concept of enterprise networking. This does not seek to distinguish who owns or controls the components that comprise the network; instead, it is defined in terms of the applications, media, customer premises equipment, public services, and operations management required to satisfy the information management and telecommunications requirements of an organisation. The focus is no longer on equipment owned, rather on how access to and processing of information is best controlled, managed and maintained (Guilder 1991).

Enterprise networks allow the sharing of information among the various parts of an organisation across its geographically dispersed locations, whether they are located in the same building or across the globe. These networks integrate the considerable computing power of the corporation for improved productivity and competitiveness.

Operationally, there is a split of enterprise networks into private local equipment and public network services (primarily to allow cost optimisation of networking equipment) but the key point is that, in the way that they work, organisations are increasingly managed as logical entities rather than physical ones. It is what they know and can find out that matters more than where they are and what they own.

So, what does this dramatic advance in capability mean?

Sure enough, there has been a terrific shift in what can be done, and in a relatively short period of time too. As stated earlier, though, it is necessary to consider social as well as technical changes to understand the likely course of future events. We now move on to look at some of the main drivers and trends that seem likely to forge the shape of information networks.

As a precursor to this, it is worth reminding ourselves of some of the more notable technical advances over the last 50 years or so. The set given in Figure 2.1 does not include the ongoing and complex technology moves in distributed

Year	Event	Impact
1944	Early computer	The dawn of non-mechanical computation
1947	The transistor	The basic building block of modern electronics
1958	Integrated circuits	Enabled powerful computers
1965	Intelsat	Basis for global telephony service
1966	PDP8 appears	Popularised computers as processing engines
1968	Optical fibre	Provided high-speed and bandwidth communications
1969	Arpanet	Early combination of communications and computing
1970	Floppy disk	Cheap information storage available
1971	Microprocessor	Cheap processing power available
1975	Ethernet	Local area networks appear
1976	PC appears	Computing power arrives on desktop
1979	Compuserve	Commercial network-based information service
1981	CD-ROM	Bulk storage medium available on desktop
1984	ISDN appears	High-speed public switched data network
1986	PCWindows	Multiple applications via computer windows
1990	PDAs, etc.	Handheld 'written input' processors appear
1995	Mosaic, Netscape	World Wide Web browsers become commonplace
1998	Communicator	Integrated phone and computers

Figure 2.1 Some notable technical landmarks, 1940–1990

computing data communications and intelligent networks, rather the more tangible products of technical advance. Even so, most people's perspective of these events is out of line with what actually happened

One lesson that can be learned from the past is that technology does not drive the real world, at least not directly. In some cases, there is a short lag between technical feasibility and common practice (e.g. the adoption of Compuserve, which took off almost as soon as it was launched). At other times, the link between feasibility and adoption has been less immediate (e.g. the use of CD-ROM has only really taken off since multimedia applications escalated local memory requirements). A secondary factor has been required to trigger action.

The way in which change has actually been brought about is complex. It requires a groundswell of either technical or social pressure to drive a potential change into practice. Even then, legal, regulatory or economic factors may advance or inhibit change. Prediction has never been an exact science, and this is one thing that will not change.

Having said this, technological advance does make it possible to do new things, and people will always been keen to exploit new ideas. The next few sections should, therefore, be taken as the necessary background to inform that exploration. Some of the points may need to be moderated against the reader's background, current position and local environment (Naisbitt 1982). But change is likely to be endemic in the information age and it is safer to treat it as a planned exercise rather than an adventure.

2.2 TRENDS AND DRIVERS

Our brief historical outline gives some perspective on the current situation, at least in terms of how technology has enabled more sophisticated operations. Along with advances in technology, we have seen ever-increasing expectations on the part of users. But what are the key factors that combine with new technology and rising user expectations to drive the future (British Computer Society 1990)?

Successful acquisition and operation of an information network will call for rigorous assessment of both technology and carrier service against the requirements of the virtual organisation, an entity reliant on its information. To build networks supporting the information infra-structure of the business, those charged with the job should take the following into account.

Scalable and enduring network architecture

The recurring costs of bandwidth and network operations and mainte-nance far outweigh capital investment in network components. Design and management are the vital enablers to the delivery of ever more complex systems (Norris, Rigby and Payne 1993).

Network bandwidth

The difficulty of estimating data traffic flows within an enterprise favours network architecture that can dynamically satisfy the bandwidth requirements of an application.

Low end-to-end latency

The above point implies the adoption of packet switching. This, in turn, means that the end-to-end delay (the sum of all transmission medium and switching fabric propagation times) must be less than the application service objective. The delays associated with small data files in the 1960s will not be tolerated for multimedia transactions in the 1990s.

Broad range of cost/performance options

Enterprise sites vary considerably in capacity and capability requirements. Also, as stated earlier, most organisations are likely to be a mix of value seekers and economy seekers. They will want to tailor their network needs to suit their information requirements (Pine 1982).

Any-to-any connectivity

The enterprise environment demands easy access to any logical location. Network names, addresses and interfaces should allow connection from anywhere. Users expect to access remotely provided services as easily as they phone distant colleagues.

Ease of installation and operation

Enterprises want to focus their energies on beating the competition, not on becoming network operators. That they have established their own telecommunications departments reflects their need to feel in control, and systems that the telecommunications service providers have not given them what they need.

Multiple types of data service

The emerging desktop environment allows the user to work with a mix of interactive data, graphics and video. Networks must support different types of data service through the same interface, sometimes simultaneously.

Increased management control of data flows

Applications vary in their network requirements with regard to sensitivity to delay. Real-time applications, such as voice and video, need predictable latency, whereas data transfers usually accept variable delays. Since a single backbone may carry both types of traffic simultaneously, the network must use mechanisms to arbitrate bandwidth access and traffic flow.

Security

The value associated with information means that organisations need to treat it as an important asset. Appropriate mechanisms for access control, user authentication and data encryption need to be included as part of an enterprise network.

Reliability

Applications availability must be the ultimate measure of network reliability. It is the product of a resilient network design, component reliability, and 'mean-time-to-respond'. Reliability issues to be included in enterprise network design include contingency planning for failure scenarios, high mean time between failure and remote troubleshooting, repair and configuration of network nodes.

We now look at a few of the key facts and figures that combine with new technology and rising user expectations to drive the future

Growth of data applications

Data traffic has overtaken voice as the driver in private networks. Figure 2.2 shows the split between voice and data traffic. The percentage of wide area bandwidth that represented data traffic rose from 25% in 1985 to 44% in 1990, and had passed 50% by the end of 1994. By 1998 the ratio is over 60% for data to under 40% for voice. Why? Because voice is saturating, and non-voice applications are growing rapidly and are demanding ever greater bit rates.

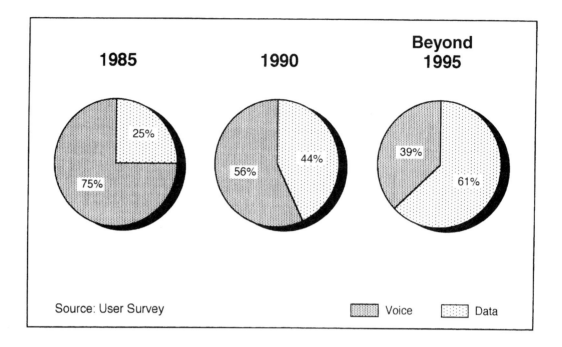

Figure 2.2 The rise and rise of data

Increasing speed

Both local and wide area network transport rates are reaching 1982 computer backplane speeds (e.g. 33 Mbps to 2.4 Gbps), blurring the line between a computer as a single site or distributed entity. This advance in speed also allows voice, video, speech, etc., all to be treated as data services, thus promoting multimedia communications (Ayre 1991).

Distribution

Centralised processing (e.g. mainframe computing) has yielded to desktop computing for many tasks, particularly those which have a real-time display orientation. The amount of power resident on the desktop, compared with that on remote host machines, has risen dramatically (see Figure 2.3). Distribution is a given fact in the Information Age.

Less predictable network traffic

Distributed computing has resulted in less predictable traffic flow than was the case for central processing. The ability to draw on services that

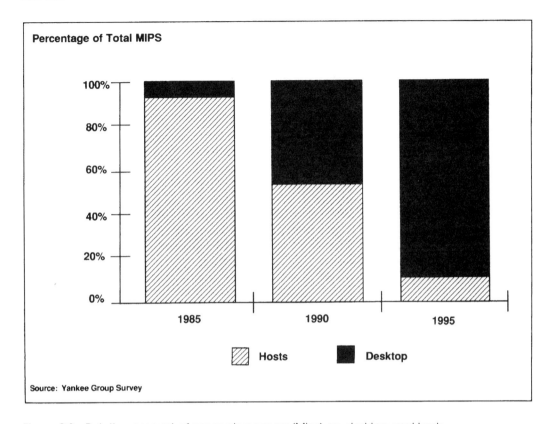

Figure 2.3   Relative amount of processing power (Mips) on desktop and hosts

reside in a wide range of physical locations means that users will generate highly non-deterministic network traffic.

Service-oriented networks

The adoption of 'standardised' (client/server) approaches to delivering applications has enabled information services to be more readily and uniformly provided over networks (see Appendix 1 for background information on client/server and other distributed computing concepts).

More flexible organisational structures

Organisations are focusing increasingly on their core business (Handy 1991). They are increasingly buying in (or outsourcing) specialist services. For information-intensive organisations, the network is the key enabler for this way of working.

Fewer boundaries

The rapid technological revolution has influenced world-wide political reform, most notably in the widespread deregulation of public telecommunication operators. This has resulted in fierce international competition, the availability of new services and the removal of many trading and physical constraints.

More network R&D

Finally, recognising the importance of advanced telecommunications, the pursuit of regional advantage has spurred new network technologies through major research and development programmes (e.g. RACE in Europe).

One further trend that is worth mentioning here is the likely growth of specialist providers of networks. The stringent requirements listed above, combined with complex technology and demanding users will push the (already established) move to outsource the provision and management of network services. The early part of the 1990s saw the outsourcing of network services grow into a billion pound business, doubling in volume every year (Lacity and Hirsheim 1993). These specialists will be asked to provide a 'Virtual Private Network' to their customer, a resource that looks like an integrated whole, despite comprising many elements from many sources.

Increasingly, network provision and operation will become a specialised business, and associated with this will be specialist information-based services. It is already the case that some organisations choose to employ independent 'information brokers' to find and collate data from a range of resources. This is likely to be but one of the information processing specialisms available in the future. A broader picture of working in the information age will be painted later in this chapter. For now, we concentrate on those whose future depends on successful Total Area Networking, the value seekers in the vanguard of the information revolution.

In order to make capital from the above information, some understanding of the state of play in both high-speed networks and distributed computing is required. As we move towards the second millennium, the challenge for those who manage information-intensive businesses will be to build and manage their network as a single entity. The traditional tasks of the network designer, such as procuring public carrier circuits faster and cheaper, will be overtaken by the need to configure complex data paths to enable access to and storage of vital information.

The first step on this path is to understand how current networks and services are likely to evolve towards Total Area Networking and Superconnectivity.

2.3 THE WORLD WIDE WEB

Some aspects of Total Area Networking are already with us (Frost and Norris 1997). Although not supported by a uniform high-speed infrastructure, the Internet (a network of computer networks that share a common set of protocols and address space) has grown over the last 25 years or so and now links together many millions of people worldwide.

Originally conceived by the US military as a means of removing reliance on a central computer, the Internet has grown into a global community of users. Most of this community use the facility to send and receive mail messages, to exchange files and to access common data. This has been enabled by the adoption of a family of protocols (known as TCP/IP) that allows most users to communicate with most remote services, irrespective of local equipment type.[1]

The Internet has done much to bring down national and organisation barriers, to enable information working, and the way of working sparked by the Internet cannot be ignored. It is already as embedded into communications culture as are the roads and oceans. It can be as readily removed, a point reinforced by the dramatic growth in Internet connections over the last few years.

This growth has prompted changes in the way that many people now use the system. The volumes of data associated with the expanded user base has meant that user access to the Internet has had to evolve in order to cope. A new form of user front end (provided to users in the form of a 'client' software package) is being used to access a host of new services (provided by 'server' software, usually on remote host computers). Together, they provide a facility known as the World Wide Web, which eases the navigation of a vast and growing information space.

Since the Internet already provides a means for any computer to communicate with any other, the World Wide Web (W3) is a global application. It can be regarded as a development of earlier page-based information services (such as Minitel provided in France), though with a number of differences.

- It can support multimedia applications, at least to a limited extent. Pages of information may contain pictures, sound and links that allow the user to navigate between pages of text simply by clicking on key words or icons.

- It is not constrained by length of page. In practice a 'page' of information can be any length, though it is customary to present a 'home page' to introduce a server and to structure subsequent information for ease of reading.

- It is readily expandable both in terms of users and the facilities that it provides. The separation of World Wide Web into 'client' and 'server' components means that it is very straightforward to add extra clients and servers to the system.

---

[1] TCP/IP provides a *lingua franca* for Local and Wide Area Networks, and provides a convergence point *en route* to Total Area Networks.

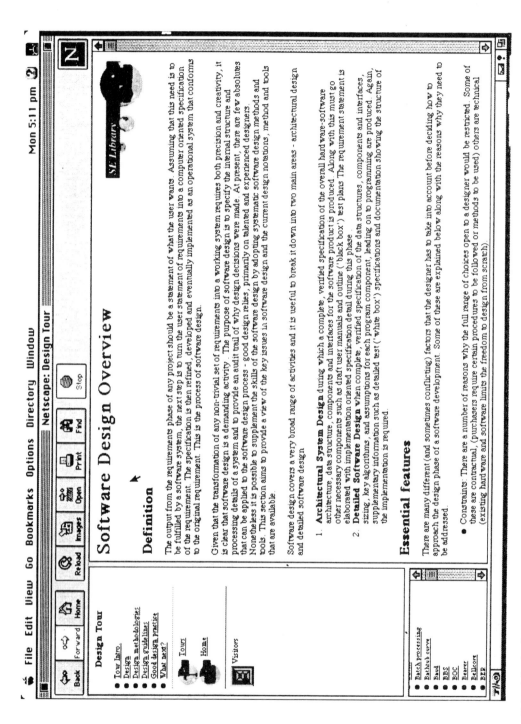

Figure 2.4 A typical page of information on the World Wide Web. (Reprinted from West and Norris, Media Engineering, 1997, by permission of John Wiley & Sons, Ltd.)

- Client and server can negotiate the document types that they will use. Typically a server offers pages based on a standard page description language but can also provide proprietary formats such as Microsoft Word.

A typical page of information from a W3 server is shown in Figure 2.4. This illustrates the mix of pictures, text, etc., and more significantly shows how the user is presented with a view of the available information (rather than just having to know where to go to find what they want).

Since the World Wide Web (W3) foreshadows much of what we expect from total area networking and superconnectivity, we will explain both its operation and technical approach in some detail.

The client software that allows a user access at virtually any type of terminal (PC, Macintosh, Unix workstation) is available from a number of sources, though the major growth of W3 was driven by the distribution of Mosaic, produced by the National Center for Supercomputing Applications (NCSA) at the University of Illinois.

Pages of information retrieved from a server are displayed in a window, much like that displayed by a word processor. Certain words or images on the page are distinguished as links, often by displaying them in a different colour, or by underlining them. Clicking on a link brings up a new page of information in the user's window. Effectively, each page can include, associated with a particular piece of hot text, some hidden text which is the reference to a different page. When the user clicks on the hot text, the client application decodes the reference and retrieves an appropriate page from the server. Although this is quite simple, there are a number of important points that should be noted.

- It is the client who decodes the reference and therefore selects the next page, which may be on any server connected to the Internet.

- The reference includes the Internet address of the target server machine (anywhere in the world); the reference to the page or document on that machine; and, optionally, a reference to a particular section in the document.

In much the same way that the postal service evolved to mask routing decisions from the user of the service, so the W3 system provides transparent access to information to its users. The focus is on information, rather than on where it resides and the routes to locate it.

As previously noted, pages may include pictures, sound or video. More recently, the adoption of the Java language has extended the choice to include applications. The client may be configured either to request these, by default, along with the text, or to load only the text and display markers where these items occur. The user is then able, selectively, to click on markers and download only those pictures, etc., that are actually wanted. This gives the user considerable control over the speed of operation and the demands on the network. In operation, anything more than text can take some time to retrieve, but the move to total area networking will both remove this restriction and (most likely) spark further developments.

To help the user navigate through the complex information that it holds, the World Wide Web employs a specific protocol (which runs over the basic Internet standard TCP/IP) called HyperText Transfer Protocol (HTTP) for the retrieval of documents by the client from the server. This follows a simple sequence:

- the client is connected to the server
- there is a simple negotiation of mutually agreeable document formats
- a document is requested
- the document is sent to the client
- the client disconnects

and it is optimised for this sort of use, transferring short documents. It effects this by being much lighter in weight than many of the standard file transfer protocols.

As hinted at the start of this section, there is significant relevant purpose in explaining the World Wide Web, as it provides a good window on the likely uses of the future network. There are differences, though, the main ones being that both the W3 and the Internet are freely provided services with many contributors (a large part of their success), that they open to anyone who can organise their own network access and that they rely to a large extent on the information providers for successful operation. (For illustration, W3 servers can be accessed at Xerox PARC, NTT, IEEE, The United Nations, Open Software Foundation, AT&T, BT, NASA and many major universities.)

Putting this another way, there is no guarantee of performance or of security. The systems operate through (distributed) goodwill and expertise, an arrangement that makes them ideal for some purposes, less so for others. Adoption to date has been fuelled by the breadth of information providers and the availability of significant desktop processing power. The connections within the Internet vary widely, with the vast majority using modems and dial-up access over telephone lines. This variability of infrastructure extends to the connections between servers (the computers on the Internet). Some have high-capacity links, others are less well heeled.

If information is vital, then the total area network will have to provide reliable, fast, managed connections. In addition, the boundary between open and private information will need to be clearly made. As part of a companys strategic resources, the network, with the applications and data that use it, will need to be protected and access-controlled. The element of commercial-isation will drive down user tolerance; assured levels of performance, reliability and security will be expected.

The final point in this section is that total area networking is now being made available internationally by a number of vendors. Enterprises can, therefore, buy enterprise networks; hitherto they could only buy the bits and pieces, which they had to assemble and manage themselves. The future

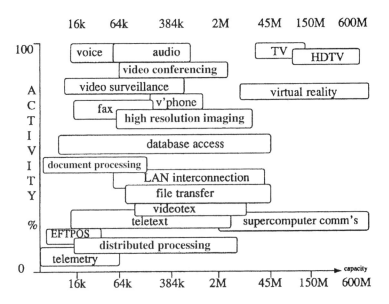

Figure 2.5a   The range of network-based applications

will see the commoditisation of Total Area Networks with resilience, reliability, performance and security as the differentiators between competing providers.

2.4   THE INFORMATION-INTENSIVE MARKETPLACE

The user pull for total area networks can be readily judged by looking at the speed with which the World Wide Web grew. In the first year of its existence, its global user base climbed to reach the 1 million mark.

Given that there is a clear demand for information networks, there are two immediate questions that need to be asked. First, what form will the information to be carried take, and second what sort of infrastructure needs to be put in place to carry it?

The answer to the first question is 'virtually without bound', as new ideas and the technology that enable it fuel each other. Figure 2.5a shows some of the services that are already viable.

The diagram shows the demands of each of these services in terms of the bandwidth required and the extent to which they use that bandwidth, termed circuit utilisation or duty cycle. (Different techniques and developments in coding techniques will cause applications to move around on this diagram.) Perhaps the most striking feature of the diagram is the almost continuous range of bandwidth and activity that is required.

In the past the arrival of each new service has been almost synonymous with the provision of a new network. On the occasions when a new service

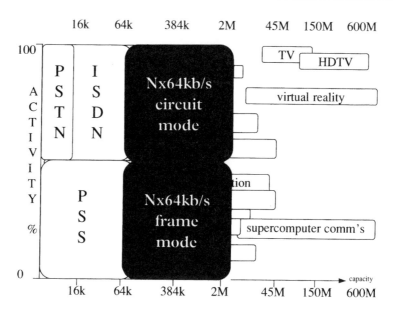

Figure 2.5b The coverage of existing networks

did use an existing network, it had to be engineered to fit that network (known as a 'derived' service). For instance, data services over the telephony network require the data to be disguised as voice by using modems.

The explosion in applications illustrated in Figure 2.5a means that it will no longer be feasible to provide different networks for different services. The way in which the 'applications map' is covered by exisiting networks (illustrated in Figure 2.5b) indicates that a new approach is called for.

The future focus will be on multiservice networks suitable for carrying information of any type. In fact, since voice, speech, pictures and video are all handled in a similar way by multimedia applications, only one network (an information network) is really called for. It is then the constraints on performance, cost and throughput that limit the application that can be supported.

The second question posed earlier (what sort of infrastructure is required) is altogether less clear cut. Given the 100 years of endeavour and investment in existing (predominantly voice) networks, the world is unlikely to change overnight. There is a legacy that cannot be ignored. Much of the rest of this book explains the component parts of an all-new, high-speed data network. Before we move on to this, though, let us take stock of where we are. Figure 2.6 illustrates the three basic categories of network support that characterise the current situation.

The first category supports voice services and derived services, including fax, that have well-defined transmission and switching characteristics. Since the public network operators have developed their voice infrastructure for over 100 years, the network typically represents the most scalable, enduring and cost-effective platform for voice traffic.

Application	Call/Message Duration	Transfer Rate (bps)	Duty Cycle	Wide Area Network Option
Voice	5 minutes	64000	20%	Circuit Switched Voice
Fax (Group 3)	3 minutes	9600	100%	
Mainframe Backup	120 minutes	2.048M	<100%	Circuit Switched Digital Transport
Video Conference	60 minutes	56k–768k	<100%	
Colour X-Ray Image	7 seconds	45M	<1%	
Database Enquiry (Terminal)	20 seconds	4800	<10%	Packet or Fast Packet Switching
Database Enquiry (Workstation)	1 second	10M	<1%	
Credit Card Authorisation	15 seconds	1200	<1%	
Check imaging	1 second	1.544M	99%	

Figure 2.6 Some of the more common applications and typical network options used to support them

A second category features applications requiring 'as needed' digital transport. This covers applications such as computer backup or high-resolution videoconferencing that mandate high bandwidth (e.g. ISDN, $n \times 64$ kbps to 2.048 Mbps) transmission facilities. For applications that have such 'casual' connectivity requirements, the number and size of simultaneous channels required and current tariffs will determine whether circuits are dedicated to each application, or if the user instead builds his own multiplexer utility backbone and provisions channels. Public network switched digital services are an attractive option where available.

The third and final category is typified by data messaging requirements that have non-deterministic bandwidth and routing requirements. Here we find applications such as electronic mail, order entry and manufacturing process control that require real-time performance to different destinations on a demand basis. Usually, this data never leaves a customer's premises (hence the rapid growth in local area networks).

Since the early 1980s, commercial LANs that interconnect workstations, servers and hosts have taken over from *ad hoc* networks composed of terminals, concentrators and mainframes. Users with data applications have had to be careful to provide a migration plan for these legacy networks. A similar evolution will have to be handled in the move towards total area networking.

The challenge for enterprise network designers will be how to collect traffic from distributed computing sources and move it to its destination whether across the room or across the world.

In the last part of this chapter we provide some of the basic guidelines on evolving technology and ways of working in the information age. The next two sections take current know-how as a basis for painting a picture of future networking. This informed extension of where we are is intended to answer the 'so what' questions that need to be answered in building a reliable route map.

2.5   VIRTUAL ORGANISATIONS

The types of facility described so far in this chapter will provide the infrastructure for distributed teams. It will be increasingly viable to have work carried out by 'virtual teams', people who work together entirely through the network (Barnes 1991). It is these non-hierarchical groups of information workers who will predominate in future business by bringing together local and specialist knowledge and focusing it on a specific goal.

The combination of global reach and local knowledge (access to fixed resources, wherever) will increasingly become a vital part of competitiveness (Naisbitt 1994). Some of the consequences of moving from central control to co-operative distribution are illustrated in Figure 2.7. The main challenge in this migration is to establish common purpose in a physically scattered organisation

This does not just happen; virtual teams have to be managed if they are to capitalise on their flexibility. This can be very difficult with so many interesting and readily available diversions to distract one from the day job. The temptation to 'surf the network' in search of the weird and intriguing (because it is certainly out there) will be great. Even if the temptation to wander in information space is overcome, there is still a challenge in focusing the diverse and distributed resources on the job in hand.

A good example of how this has been achieved in the past lies with the designers of modern aircraft. The latest European Fighter had its wings developed in one location, its fuselage in another and its engines at a third. Physical assembly was carried out as a final exercise and there was reasonable expectation that the various components would fit. This confidence was borne of experience with that way of working and of the shared understanding across the distributed team. In this instance, the end objective was tangible,

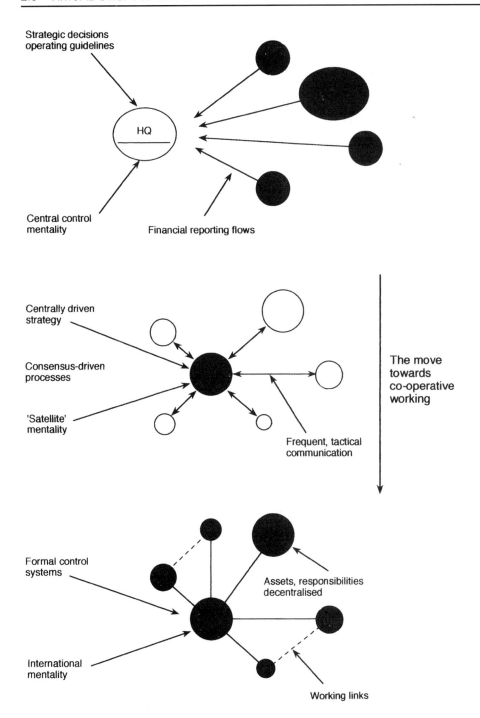

Figure 2.7   Organisational shift from central control to co-operative work

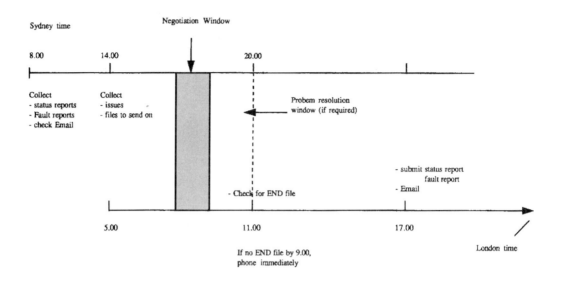

Figure 2.8   Co-operative working across time zones

something that everyone could relate to and could describe in a commonly accepted language (i.e. diagrams, component part, etc.).

The extension of this scenario (enabled by Total Area Networking) is that products can be built as logical items, pieces of electronic data, for assembly by an end user. There is an added complication here, though, when compared with the example above. Abstract tasks, such as software development, do not readily lend themselves to the sort of measurement or visualisation as with aeroplanes. Successful distributed teams need to be carefully organised and not simply equipped with relevant technology if they are to function effectively.

To exploit virtual teams in the broadest sense, it will be necessary to implement procedures to monitor, control and synchronise distributed team efforts. With the comfort factors of personal contact and/or shared experience taken away, the value this provides needs to be replaced (Quarterman 1989). Certainly, it is the author's experience that virtual teams are made, not born, and that even the best enabling technology will only pay dividends if matched to the people who use it.

For instance, individuals must be able to ascertain precisely what pieces of work are complete, which are still in progress and which have yet to be tackled. Also, they usually benefit from some idea of what a distant coworker is likely to deliver (and when). With geographical separation often removing the option of asking the relevant person (it may be the middle of their night when you want to know), some overhead to supplant traditional methods needs to be put in place.

Exactly what has to be added in terms of process depends on the type of work that is being done. This is illustrated in Figure 2.8, which outlines some guidelines that have proved useful in the planning of distributed software developments.[1]

This sort of strict schedule for information transfer, checking and problem resolution is but one part of effective co-operation. In general, there are a number of key factors that need to be addressed as operations shift from the co-located to the globally separated (Lu and Farrell 1989), some concerned with control, the majority reliant on management borne of understanding. These issues are detailed in Appendix 2.

2.6 PARTS OF THE JIGSAW

Combine the World Wide Web in its current state with the effectiveness of existing virtual teams and trends and drivers outlined earlier and you find some user needs as yet unfulfilled. Basically, there is a model of the world to come, but it is not yet sufficiently developed to allow the full potential to be exploited nor secure enough to be trusted with critical data. It is in these areas that the next steps will come, virtual private networks (VPNs) as managed and controlled parts of a wider information network (the volume retailers on the high street of the information cities). They will need inter-city highways as a core part of their business! We now introduce the key technical elements of total area networks. Each will be covered in some depth later; for now, a brief introduction.

As the use of personal computers continues to grow, so does the need to connect them to share applications and resources. Many companies today need to expand connectivity cost effectively beyond their local area networks (LANs) to exchange critical information between remote locations. To do this requires high-bandwidth low-delay connections spread across a wide geographical base. We now have a brief look at the new network technologies that will allow this to happen. Subsequent chapters deal with each in much greater detail; for now, we only aim to outline the major characteristics of each and position them, one against another.

Frame Relay

Frame Relay is a protocol that allows interconnection of LANs and their associated applications. A reasonable summary of Frame Relay would be that it is the lightweight successor to X.25, the long-established base for packet switched data networks. The greater availability of digital lines has made Frame Relay an attractive protocol for the transmission of large amounts of data. The low level of error checking, etc., designed into Frame Relay makes it faster than X.25 with little or no performance degradation. Frame Relay technology is well suited for intermittent or 'bursty' traffic characteristics that

---

[1] In practice, attention needs to be paid to operational dynamics. In this example UK summer time is shown. Adjustment is needed to cope with the (asynchronous) transition to winter time.

require high network bandwidth on demand in order to meet demanding response-time requirements.

Frame Relay offers a high-speed low-overhead protocol standard that can be used over a wide area network (WAN). Because the Frame Relay protocol takes advantage of the widespread availability of error-free digital lines it delivers high speeds, protocol transparency and low propagation delay similarly to leased lines, but at a lower cost.

Using Frame Relay, the grade of service for each PVC can be controlled through a feature known as committed information rate (CIR). This can be set (usually in 16 kbps increments) to suit application requirements and link capacity. The CIR allows an assured level of throughput, a baseline that the user can depend upon.

If the user exceeds the predetermined CIR on a particular circuit, the additional information may still be delivered, but some frames may be dropped. This loss of frames is not something that the Frame Relay protocol can do anything about; it is left to higher-level end-to-end protocols to recognise this and to trigger the re-sending of information.

There are obviously refinements, but the simple picture given above captures the essence of Frame Relay, a simple, fast protocol that provides a flexible platforms for building information networks over reliable digital links. By the early 1990s, a number of vendors were offering a Frame Relay service, and this is now available world-wide.

SMDS

The switched multi-megabit data service (SMDS) was the brainchild of Bellcore, the research organisation for the US Regional Bell Holding Companies. Its mission was to establish an affordable method to enable its client companies to interconnect LANs over a wide area, much the same application area as described above for Frame Relay.

Four key service characteristics have been defined within SMDS: a connectionless service with multicast capabilities; global addressing to enable intercompany transmission; a facility for creating virtual private networks; and broadband performance that protects users from having to pay for unused bandwidth.

In Britain's academic circles SMDS is already being put through the mill in one of the most bit-hungry environments of all—scientific computing. SuperJANET (the super joint academic network) connects more than 50 university campuses over SMDS links, providing at least 70 times the performance of the preceding privately-owned network.

Since both SMDS and Frame Relay fulfil the same basic function, that of interconnecting LANs, the choice of which to use is one of cost and performance. In terms of tariffs, the two are usually fairly close: there are examples of countries where SMDS is cheaper and of where Frame Relay is more competitive. On purely technical grounds, the higher overheads of

SMDS may limit its long-term uptake. This is likely to be particularly true for intelligent networks, where the need for routing decisions on a per packet basis will militate against connectionless operation.

ATM

Asynchronous transfer mode (ATM) provides a basis for multiservice networks, and it can carry both Frame Relay and SMDS. The basic operation of ATM is to route short packets (called cells) at extremely high rates. Cells, which contain up to 48 octets (an octet is eight bits) of data and five octets of addressing and control information, can carry digitised voice, arbitrary data and even digitised video streams.

Conceived, originally, as part of a broadband ISDN (BISDN) system design, ATM switches along with fibre transmission technology are likely to provide the 21st century equivalent of the 20th century telephone network. BISDN offers the promise of a common network for all information and communications services, rather than special networks for different services, such as voice, data and video.

Asynchronous transfer mode (ATM) is the technology that will deliver true multimedia to the desktop. The arrival of ATM as a commercial reality heralds the start of the information age, based on the multimedia communications that it enables. Some large corporations are already installing ATM in LANs, thereby redrawing the boundary between their own networks and the public network, capturing more and more of the functionality and added value to their side.

With these technologies in place, giving an infrastructure to link a group of physically disparate people, whole new working scenarios emerge (IEEE 1992). There follows some examples.

- Group authoring: experts, wherever they may be located, contribute to a document held on a remote machine. Review, editing and formatting of the document is carried out during production and the draft is loaded onto a newsgroup so that all involved can see what is going on. Pictures, voice notes and associated video clips can be added as appropriate, both on- and off-line.

- Remote monitoring: the performance measures of a remote piece of equipment are collected for a remote operator to effect required changes. This sort of application is already in place for adjusting turbines when they first go into service and for network maintenance and repair (not all shoemakers go unshod!)

- Follow-the-sun development: where the development of, for instance, software is divided into time slices to suit the normal working hours around the globe. As people shut up shop for the day in Tokyo, their working files are copied to Bombay for further development and at the close of the Indian day, to Frankfurt for test. Next day in Tokyo, the team find an extended, tested code. As well as following the sun, this type of operation can also be an effective way of getting the right mix of skills.

- On-line Shopping, catalogues: people sit in their offices or their armchairs to buy goods rather than go to the shops. The ability to present high-quality pseudo-goods on-screen and to allow the users to select what they want, perhaps customise it and then pay for it, will fuel the development of 'transaction trading'.

- Remote surgery: where the operations in one location are guided by an expert (or team of experts) in another! This may seem a little worrying, but there are already instances of effective 'telepresence' applications for medical applications.

- Home Services: where the facilities currently available on interactive compact disk (shopping catalogues, movies, games) are available 'live' over the network, rather than 'canned' on a disk.

These applications all rely on a network that is fast and reliable enough to allow people to work as if on the same local network, even though they are some distance apart. Once the tyranny of distance has been removed, new ways of using and exploiting total area networks may emerge, just as new ideas have been sparked by the Internet and local area networks (Guilder 1993). In its turn, this will fuel further network advances, such as embedded intelligence, transaction trading (a network-based capability that allows a service to be automatically requested, fulfilled and paid for) and the like (see Appendix 3 for an overview of the likely evolution of public networks).

The pace of application development to exploit a multiservice network has yet to come clear. There is some clue to this in the rapid growth of Multimedia—coding techniques such as MPEG, with matching tools to use information have very quickly popularised usage. Control over the vast information space afforded by globally distributed computers is another story, though (Simon 1992). In this book we stick to the highways, not the way they are used. Having said that, it is worth considering the shape of things to come.

One notable effect of the Industrial Revolution was that it provoked a move from village to town life. There is likely to be a move in the reverse direction with the current Information Revolution. You will be able to work where you like (physically) and still be part of a large 'community'. The distance will have been taken out of information.

The information villages (a term coined by Marshall McLuhan) seem likely to evolve into high-tech towns once the highways that connect them are put in place.

To extend the analogy, the small information corner shop (already on the World Wide Web) will persist but will be joined by information banks, retailers and factories (courtesy of the value-seeking organisations). There will also be a range of new facilities, ranging from the information pub, where salacious gossip is exchanged (much like some of the racier user groups on the Internet) to the community doctor for the sick of the information city.

The Information Age is still in its infancy, though. There are certainly thriving information villages but the towns have yet to appear; they will depend on the highways. The next chapter starts to explain how the current linking paths will evolve into these highways that support the migration.

2.7 CONVERGENCE AND COLLISION REVISITED

In the previous chapter we explained how the delivery of the Total Area Network relies on the convergence of computer and telecommunication technologies, yet is hampered by the very different approaches they take. Since then, we have focused on convergence. To redress the balance, and more accurately reflect reality, we now reintroduce the collision aspect.

In doing this, we put forward a definition of a Total Area Network as any communications platform where distance is not an issue. Within this definition there are clearly many different options. Some issues pervade all of the options, for instance the need for end to end design, the configuration of local and wide area network elements to provide a service, combined service and network management, and so on.

There are, however, some issues that depend on the technology chosen to build the Total Area Network. For instance, with a shared medium network such as the Internet, security is a key issue. Scalability on the other hand is almost a built-in attribute. By way of contrast, in a traditional telecommunications network with a hierarchy of switched paths, security can readily be assured but scalability needs to be carefully planned.

This contrast is more than one of a guaranteed service network against a best efforts data network; it typifies the different approaches taken by the leading players in future information networks. An example of this is that the operators of national telecommunications networks invariably feel (or are obliged to ensure) that their networks and service are closely monitored and managed. Many Internet providers, on the other hand, are content to provide a service for which the end user takes significant responsibility.

For the next three chapters, we park the different flavours of Total Area Networking and concentrate on technology. Then, in Chapters 6 and 7, respectively, we return to the points raised here and illustrate how the telecommunications and computing communities go about building their version of the Total Area Network. Not that one is right and the other wrong. Simply that they are different and an appreciation of those differences is important in getting the flavour you prefer.

2.8 SUMMARY

In many respects we know what the network for the Information Age will look like. We know that

- the infrastructure to make it 'industrial strength' is available

- it will be driven by value seekers with economy seekers coming on board later

- it will push new ways of working and new organisational structures, some of which we already know about.

To reach this conclusion, this chapter has explored the various trends and drivers that seem destined to shape the Information Age. The impact of these influences (social as well as technical) are used to point the way ahead and to identify the landmarks on the route ahead.

Having drawn the broad picture of the future, the enabling foundations of Frame Relay, SMDS and ATM are introduced, precursors to the more detailed treatment that follows.

REFERENCES

Ayre, J. (1991) *A Beginner's Guide to Multimedia.* UK National Interactive Video Centre.

Barnes, I. (1991) Post Fordist people. *Futures*, November.

British Computer Society (1990) *The Future Impact of Information Technology.* BCS Trends in IT series.

Frost, A. and Norris, M. (1997) *Exploiting the Internet.* John Wiley & Sons.

Guilder, G. (1991) Into the Telecosm. *Harvard Business Review*, Mar–Apr.

Guilder, G. (1993) When bandwidth is free. *Wired*, Sept–Oct.

Handy, C. (1991) *The Age of Unreason.* Business Books.

Karimi, J. and Konsynski, B. R. (1991) Globalisation and information management strategies. *Journal of Management Information Systems*, **7**, No. 4.

Lacity, M. and Hirsheim, R. (1993) *Information Systems Outsourcing—Myths, Metaphors and Realities.* John Wiley & Sons.

Lu, M. and Farrell, C. (1989) Software development—an international perspective. *Journal of Systems and Software*, **9**.

Monk, P. (1989) *Technological Change in the Information Economy.* Pinter Press.

Naisbitt, J. (1982) *Megatrends—Ten New Directions Transforming Our Lives.* Warner Books.

Naisbitt, J. (1994) *Global Paradox.* Nicholas Brealey Publishing.

Norris, M., Rigby, P. and Payne, M. (1993) *The Healthy Software Project.* John Wiley & Sons.

Pine, B. J. (1992) *Mass Customisation—the New Frontier in Business Competition.* Harvard Business School Press.

Quarterman, J. (1989) *The Matrix: Competitive Networks and Conferencing Systems Worldwide.* Digital Press.

Simon, A. (1992) *Enterprise Computing.* Bantam Professional Books.

Strategic impact of brandband communications in insurance, publishing and healthcare. *IEEE Journal on Selected Areas in Communications*, **10**, December.

In addition to the 'hard' references given here, there is a huge amount of information relevant to the total area network to be found on the network itself. The best way to find what you want is to explore what is there, but a couple of useful starting points are mosaic@ncsa.uiuc.edu and the user group news.announce.newusers. The former is a way into the World Wide Web, which carries on-line guides such as 'The Internet Companion', the latter provides a range of bulletins that answer frequently asked questions (FAQs).

For those who have no network access, a useful contact is the Internet Network Information Center (telephone +1-619-455-4600). If you have some form of network access (readily obtained from many network operators, universities or service providers), there is an automated mail service from this source. It can be accessed as follows.

- Mail to mailserv@is.internic.net
  (with message: begin
                 help
                 index
                 send about-information-services/contact-info
                 end)

or

- telnet gopher.internic.net
  (type 'gopher' at the login prompt. It is also interesting to telnet other public servers such as gopher.uiuc.edu, archie.rutgers.edu and info.cern.ch)

or

- ftp is.internic.net
  (type 'anonymous' at the login prompt and then follow screen instructions).

3
Frame Relay

Systems should be as simple as possible but not simpler

Albert Einstein

Many people are familiar with the first widely used service of the datacomms era—X.25. As a mature product of the 'analogue era', it is a complex protocol. Much of this complexity arose from the need to protect against errors introduced by noisy analogue transmission circuits and the comparatively long round-trip delay caused by their low transmission speeds. Frame Relay, by contrast, is very much a product of the 'digital age', exploiting the much lower error rates and higher transmission speeds of modern digital systems. In particular, Frame Relay has its roots in the Integrated Services Digital Network, the ISDN (Griffiths 1992).

3.1 THE ISDN

The ISDN is the latest step in the evolution of the PSTN. The early 1970s saw the widespread introduction of digital transmission into the telephone network. This was followed in the late 1970s by digital switching systems with computer control of switching operations together with powerful message-based inter-processor signalling between the switching centres. This combination of digital switching and digital transmission systems—the so-called Integrated Digital Network or IDN—promised great flexibility for new service innovation and reductions in operating costs. But the local access circuit between the user and the local exchange was still analogue. It was still just a telephone network.

Driven by the increasing importance of non-voice services, the 1980s saw the next logical step in the evolution of the telephone network in which the digital connection is taken all the way to the user making it equally suitable for voice and non-voice services. To exploit the power of this all-digital

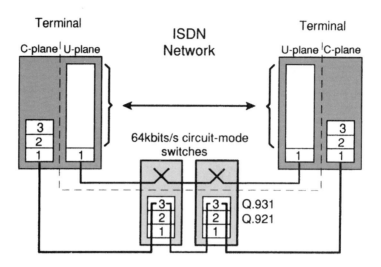

Figure 3.1 Separation of user information and signalling in the ISDN

network the message-based signalling was similarly extended all the way to the user. This development creates the Integrated Services Digital Network, the ISDN.

As Figure 3.1 shows, one of the distinctive features of the ISDN is that the user information and signalling are kept logically separate from end-to-end through the network. In ISDN parlance, the user's information lies in the user-plane (U-plane) and the signalling lies in the control-plane (C-plane). It can be seen that in effect the ISDN is composed of two sub-nets, a switched information sub-net (above the dotted line) and a signalling sub-net (below the dotted line).

The signalling protocols are layered in accordance with the OSI reference model. There is a network layer call control protocol, specified in ITU-T standard Q.931, which defines the call control messages and procedures (types, formats, meanings, interactions and so on). There is also a link layer protocol, defined in Q.921, which makes sure that the call control messages are reliably passed, without errors, between the terminal and the call control process in the serving local switch.

Basically the call set-up procedure illustrated in Figure 3.2 is as follows. A terminal initiates a call by creating a SETUP message and sending it to the serving local switch. This message contains the calling and called line addresses, as E.164 numbers, together with any other information needed to establish an appropriate connection such as terminal compatibility information; there is no point in trying to set up a connection between a fax machine and a telephone, for example!

After performing the usual validity checks (does the SETUP message contain all the information needed for call establishment? was the last bill paid in time?, etc.) the local switch acknowledges receipt of the SETUP message by returning a CALL PROCEEDING message that indicates that the call is now being set up. It then routes the SETUP message to the next

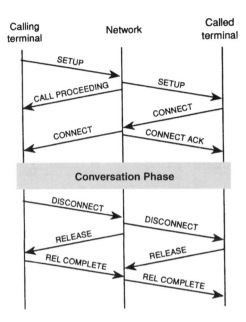

Figure 3.2 Basic call control using ISDN signalling

exchange *en route* until it reaches the destination local switch, where it is passed to the called terminal.

On receipt of the SETUP message the called terminal may accept the call by returning a CONNECT message as shown, causing the switch cross-points to be switched *en route*. Finally the destination local exchange sends a CONNECT ACKnowledge message to the called terminal, indicating which of the available channels will carry the call. The call enters the 'conversation' phase.

At some later time a corresponding exchange of signalling messages clears the call as shown. For a more comprehensive and rigorous account of all aspects of the ISDN the reader is referred to the companion volume *ISDN Explained* (Griffiths 1992).

As a development of the PSTN, the ISDN is intrinsically circuit-switched. But for many data applications packet-switching is clearly more appropriate. X.25, however, does not fit the ISDN model of keeping user information and signalling separate. Nor is it necessary to include X.25's heavyweight error correction protocols in the comparatively error-free digital environment. Something else was clearly needed in the ISDN to support data services effectively and Frame Relay was defined to fill this gap.

3.2 FRAME RELAY AS AN ISDN BEARER SERVICE

Frame Relay is a simple connection-oriented, virtual circuit packet service. It provides both switched virtual connections (SVCs) and permanent virtual

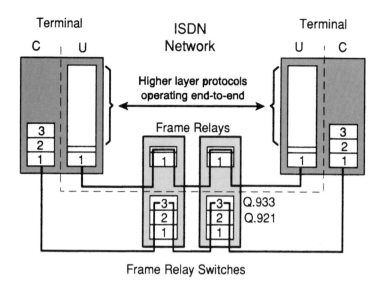

Figure 3.3 ISDN Frame Relay

circuits (PVCs), and it follows the ISDN principle of keeping user data and signalling separate.

An ISDN Frame Relay SVC would be set up in exactly the same way as an ordinary circuit-mode connection using ISDN common-channel signalling protocols as outlined above. The difference is that in the data transfer, or 'conversation', phase the user's information is switched through simple packet switches (known as frame relays) as shown in Figure 3.3, rather than circuit-mode cross-points.

PVCs would of course be set up on subscription by the network operator; user signalling is neither needed nor provided. To cover the additional call parameters and procedures needed for frame mode services, an enhanced version of Q.931 (the call control signalling protocol) has been defined, known as Q.933.

The Frame Relay data transfer protocol

In Frame Relay information is transferred in variable-length frames with the simple format shown in Figure 3.4. In addition to the user's information there is a header and trailer, each of two octets.[1] The header contains a ten-bit label agreed between the terminal and the network at call set-up time (or at subscription time if a PVC) which uniquely identifies the virtual call. This label is known as the Data Link Connection Identifier (DLCI).

Terminals can therefore support many simultaneous virtual calls to different destinations, or even a mixture of SVCs and PVCs, using the DLCI to

---

[1] The header is normally of two octets, but the standards allow for three and four octet headers which can accommodate longer labels.

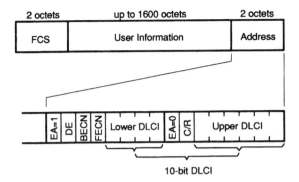

Figure 3.4 Format of Frame Relay frame

identify which virtual connection each frame belongs to. DLCI values 16 to 991 are available to identify the user's SVCs and PVCs. Other DLCI values are reserved for specific purposes. For example, DLCI = 0 is used to carry call-control signalling, and DLCIs 992 to 1007 are used to carry link layer management information as described below.

HDLC flags (the bit pattern 01111110) are used to indicate the beginning and end of each frame and as interframe channel fill, with zero-bit insertion and deletion used to avoid flag simulation in the user information field. This is exactly as in X.25. The minimum amount of user information that a frame may contain is one octet, and the default maximum size of the information field is 260 octets. However, most implementations support up to 1600 octets to minimise the need to segment and reassemble LAN packets for transport over a Frame Relay network.

The trailer contains a two-octet Frame Check Sequence (FCS) calculated in the same way as for an X.25 frame.

The EA bit indicates address extension and follows standard HDLC practice. Set to 0 it means that another octet of address follows this one. Set to 1 it means that this is the last octet of the address. C/R, the command/response bit, is passed transparently from one terminal to the other.

A link layer protocol has been defined for frame mode bearer services. Usually referred to as LAPF or Q.922, it is based on Q.921 the link layer protocol used to carry user signalling (see Figure 3.1). The data transfer protocol used in Frame Relaying is a (small) subset of LAPF, known as the data link core protocol.

The LAPF core protocol provides for the sequence-preserving bi-directional transport of frames between terminals. It includes the detection of frame errors, but not error correction. Nor does the network operate flow control. It is left to the higher-layer protocols operating directly between the terminals to look after error correction and flow control. There is thus very little processing of frames by the network nodes, and frames can pass through the network quickly and transparently.

Figure 3.5 illustrates how simple the Frame Relay data transfer protocol really is. Consider that we have a Frame Relay terminal connected to port **x** and that we have a single virtual connection established. Remember that at call set-up time (or at subscription time if a PVC) we agreed that we would

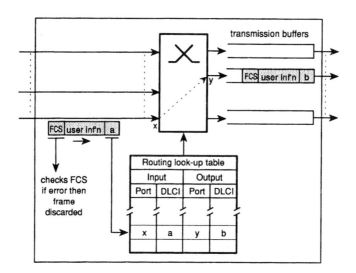

Figure 3.5 The principle of Frame Relaying

use a particular 10-bit label (DLCI) with the network. This is shown as **a**. The terminal sends a sequence of frames into the network. Figure 3.5 shows one of them: it contains DLCI = **a** in the header.

When the Frame Relay switch receives this frame it does a few checks. Firstly it looks for transmission errors using the 2-octet frame check sequence (FCS) contained in the trailer. If the frame has any transmission errors it is simply discarded. If not, a few other checks are done: is the frame too long? too short? has the DLCI = **a** been allocated? Again, if an error is found the frame is simply discarded.

Assuming that the frame gets through these checks, the Frame Relay switch then looks in the routing look-up table to see which outgoing link it should be transmitted on. Looking down the routing table for port **x** the switch finds that frames with DLCI = **a** should be routed out on port y and should be given the new label DLCI = **b** on the outgoing link. Because the DLCI is changed it is necessary to recalculate the frame check sequence before transmitting the frame.

At call set-up time entries in routing look-up tables were made in all Frame Relay switches *en route*, exactly like that shown. So our frame passes through the network until it is finally passed to the destination terminal, taking a different value for the DLCI for each link on the way.

Figure 3.5 shows only one direction of transmission. Transmission in the other direction is achieved in exactly the same way. Indeed, as we will see, one of the merits of Frame Relay is that the two directions of transmission are treated independently and can be configured to have different throughput. If we should want to set up additional virtual connections they would be given different DLCIs and the switches would process and route them independently.

The above illustration of the Frame Relay principle is intended to give a clear description of the data transfer protocol. The reader should realise that real network implementations may actually be quite different to the picture

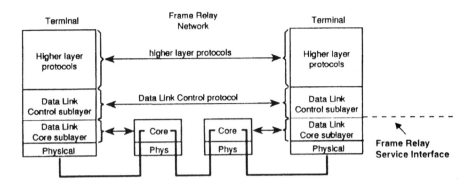

Figure 3.6 The Frame Relay protocol stack

painted here. We will see in Chapter 5 for example that an ATM network can support the Frame Relay service very effectively; and in practice implementations include functionality not mentioned here, such as the ability to dynamically reconfigure routing tables to route around link or switch failures.

But the illustration shows several things. Firstly, DLCIs have only 'local' significance. They will in general be different at each end of a virtual connection; if they are the same it is just coincidence. Secondly, the network does not operate any flow control (we will see later, however, that restrictions are placed on the rate at which a terminal may send frames). Thirdly, the network does not attempt to correct any errors it may find. Error correction and flow control are left to higher-layer protocols operating end-to-end between the terminals.

Summarising then, the data transfer protocol used in Frame Relay (the LAPF core protocol) does the following things

- identifies the beginnings and ends of frames using HDLC flags

- uses zero bit insertion and extraction to prevent the flag sequence being simulated within a frame

- detects transmission errors using the frame check sequence

- checks that frame length and, where possible, parameters in frames are valid

- multiplexing and demultiplexing frames for different virtual connections using the DLCIs

- congestion control (we will see something of how this is done in the next section).

Very few protocols can be described so succinctly!

The Frame Relay protocol stack (Figure 3.6) shows the data link layer protocol divided into the data link core sublayer and the data link control sublayer. (For simplicity signalling is omitted). The Frame Relay service is concerned only with the core sublayer. Users may choose any Control sublayer they wish, providing that it is compatible with the Q.922 core sublayer, including of course the Q.922 control sublayer.

Congestion

One of the great merits of the simple data transfer protocol is that it provides a high degree of transparency to the higher-layer protocols that are carried. This contrasts with X.25, where the scope for destructive interference with higher layer protocols often causes problems and can seriously impair performance and throughput.

But simplicity has its price. The absence of flow control leaves the network open to congestion. Congestion ultimately means throwing frames, away. Throwing frames away causes higher layer protocols to retransmit lost frames, which further feeds the congestion, leading to the possible collapse of the network. Congestion management is therefore an important issue for the network designer and network operator if these serious congestion effects are to be controlled and, preferably, avoided.

Congestion management includes dimensioning the network so that it can carry the expected traffic. It also includes implementing real-time controls in the network, which attempt to minimise the likelihood of congestion arising, recover gracefully from any congestion that does actually occur, and spread the effects of any congestion 'fairly' over all affected users. It would clearly be unfair, for example, to penalise users who are keeping within their agreed traffic profiles (see below), at the expense of more profligate users who are not.

Congestion management is not standardised. It is left to the network operators and differs from one network to another, depending on the capabilities and features designed into the switches, the network topology and dimensioning rules used, the services actually delivered to users, the control the network operator has over the CPE, and so on. But the standards do include mechanisms for indicating the onset of congestion and guidance on how CPE should respond to such notification.

Looking at the frame header in more detail (Figure 3.4), we can see that two of the bits, designated the forward explicit congestion notification (FECN) and backward explicit congestion notification (BECN) bits, are used to carry congestion indications to users' terminals. When the onset of congestion is detected by a frame relay switch, typically by a transmission queue length exceeding a preset threshold, it sets the FECN and BECN bits in the headers of any frames currently passing through the switch. As shown in Figure 3.7, the FECN bit is set in frames going towards the receiving terminal which can then use higher-layer protocols to make the transmitting terminal reduce its sending rate, typically by reducing a window size. The BECN bit is set in frames going back to the sending terminal, and achieves the same effect more directly.

Alternatively, a congested switch can send congestion notification to switches at the edge of the network 'in bulk' using something called a consolidated link layer management (CLLM) message. The CLLM message is sent on a Layer 2 management connection (DLCI = 1007) and contains a list of the DLCIs of virtual connections that are currently affected by congestion. The edge node can then take appropriate action to temporarily throttle the input of frames to the network, using either FECN/BECN or further CLLM

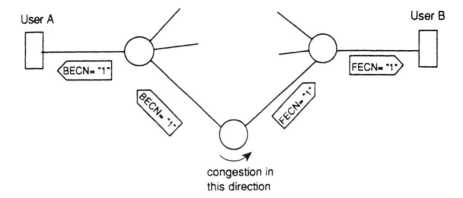

Figure 3.7 FECN and BECN indicate congestion

messages to notify congestion to the relevant terminals.

In addition to these explicit indications of congestion the terminal can, of course, sense congestion from the loss of frames or a significant increase in cross-network delay. This is sometimes referred to as 'implicit' congestion notification.

It is clearly desirable for terminals to co-operate in controlling congestion by reducing their demands on the network when notified of congestion, either explicitly or implicitly, and standards have defined procedures that should be used for this (I.370). But for obvious reasons the network cannot rely entirely on users actively co-operating in this, and must include congestion control mechanisms that prevent catastrophic collapse of the network.

In addition to the explicit congestion notification bits, a third bit, designated the discard eligible (DE) bit, can be set either by the user or the network to indicate that the associated frame should, in the event of congestion, be discarded in preference to frames in which the DE bit is not set. We will see the DE bit again.

Quality of service—what the customer actually gets

It is important for the users and the network operator to agree on the nature and quality of the service to be provided. This gives the service provider an estimate of the traffic to be expected, essential to dimension the network properly, and it gives users defined levels of service which they can select from to best match their requirements. It also gives users defined expectations against which they can complain if the service actually achieved falls short.

The Frame Relay standards specify more than a dozen parameters which characterise service quality, some relating to the demand the user will place on the network, others specifying the performance targets the network operator is expected to meet.

The key parameters that characterise the Frame Relay service are the

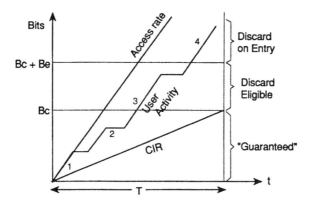

Figure 3.8 Frame Relay service parameters

committed information rate (CIR), sometimes also known as throughput, the committed burst size (B_c), and the excess burst size (B_e), all defined in relation to an averaging period T_c, normally calculated as B_c/CIR. They are negotiated at call set-up time in the SETUP message requesting the connection (or set at subscription time for a PVC). The relationship between them is illustrated in Figure 3.8.

The precise meanings of these parameters is open to a bit of interpretation. But B_c is usually regarded as the maximum amount of data the network is prepared to accept during T_c with any real guarantee of delivery; CIR is the corresponding data rate. B_e is usually taken to indicate the maximum amount of data during an interval T_c, over and above B_c, that the network will accept: this 'excess' data is usually carried on a 'best efforts' basis.

These service parameters are policed at the point of entry to the network, and they can be set independently for each direction of transmission to cater efficiently for applications that send more information in one direction than the other, such as interactive screen-based applications.

Figure 3.8 illustrates three regions of operation, shown as 'guaranteed', 'discard eligible', and 'discard on entry'. In the 'guaranteed' region (i.e. for frames 1 and 2) the network operator aims to offer a high level of assurance that frames will be delivered and dimensions the network accordingly. In the 'discard eligible' region the network will accept the traffic (i.e. frame 3) but set the DE bit: in the (hopefully unlikely) event that congestion is encountered this frame will be discarded before frames in which the DE bit is not set. In the 'discard on entry' region the frames are discarded on entry as a means of protecting the network from traffic levels likely to cause congestion.

In practice CIR on a 2048 kbit/s access circuit would typically be selectable up to a maximum of 1024 kbit/s for each virtual connection in steps of 16 kbit/s.

It can be seen that these parameters can be varied to achieve a very wide range of service levels. A typical example would be a 'committed' or 'assured' service in which network resources are strictly dimensioned, or even reserved, on the basis of the CIRs. This would give a very low probability of frame loss, provided the user did not significantly exceed his contracted CIR. Such a service could even be used to emulate a real circuit in order to carry, for example, a video application. An alternative 'statistical' service could allow the user to burst substantially above the CIR, perhaps even having a contracted CIR = 0, allowing the user to send larger bursts of data but at the risk of significantly higher frame loss rates. You pays your money and you takes your choice!

Frame Switching

For the sake of completeness the reader should be aware that a second ISDN frame mode bearer service has been defined, known as Frame Switching (I.233.2). Like Frame Relay, it follows the ISDN principle of keeping the user's information separate from call-control signalling, and a Frame Switching SVC would be set up using ISDN common-channel signalling, exactly as for a Frame Relay SVC. But Frame Switching differs from Frame Relay in the protocol used to transport frames between the terminals in the data transfer phase.

In Frame Switching the full LAPF link layer protocol is supported for data transfer, not just the LAPF core functions. This includes error-correction-by-retransmission and explicit flow control based on a sliding window mechanism. In effect Frame Switching provides the same robust data transfer capability as X.25, but, like Frame Relay, carries out multiplexing of simultaneous virtual connections at layer 2 rather than at layer 3.

At the time of writing, however, there is little or no interest in Frame Switching. It would seem that its poorer transparency to higher-layer protocols and its higher frame processing overheads outweigh any advantages.

3.3   THE FRAME RELAY DATA TRANSMISSION SERVICE

One of the dominant trends of the last decade has been the rise and rise of the personal computer. From its early use as a word processor it has evolved to become an indispensable part of a company's information infrastructure, and is now almost as common in the office as the telephone. One of the most remarkable features of this evolution has been the associated (almost explosive) growth in local area networks (LANs) used to interconnect them.

By their very nature LANs are capable of only limited geographical coverage. Companies with LANs in different locations therefore need to interwork them, often over long distances and sometimes internationally, and a confusion of bridges, routers, brouters, hubs, gateways and other devices has been developed to adapt the profusion of proprietary LAN protocols to the available wide area channels used to interconnect them.

Until Frame Relay came along, the choice for wide area LAN-interconnection lay between leased lines and X.25. Leased lines tend to be expensive, especially for international interconnection, and are not well matched to the bursty nature of LAN traffic. X.25 is a complex protocol which tends to interfere destructively with any higher-layer protocols being carried, usually degrading throughput seriously, often severely, and occasionally fatally.

Frame Relay's high speed and transparency to higher-layer protocols make it an almost ideal choice for interconnecting LANs over wide areas. This was recognised very early by a few enterprising companies who were quick to see its potential. Their vision of Frame Relay did not involve the ISDN at all. They saw it (indeed, still see it) as a wide area data networking service in its own right. This non-ISDN Frame Relay service has come to be known as the Frame Relay Data Transmission Service (X.36).

Arising out of informal meetings to resolve issues of interworking and compatibility (and not uncommonly non-interworking and incompatibility) the Frame Relay Forum was created early in 1991. Although not itself a standardisation body, the Frame Relay Forum has developed a number of Implementation Agreements which provide industry standards defining which options in international standards should be implemented and how. The Forum also recommends methods of testing and certification for conformance and interoperability, and promotes the market for Frame Relay products and services.

Stimulated by the Frame Relay Forum, the use of non-ISDN Frame Relay for LAN interconnection has developed quickly and, in addition to private networks, a growing number of network operators are offering Frame Relay services based on public high-speed data networks. Indeed, the wheel has now turned full circle and the activity and interest in non-ISDN Frame Relay has caused ITU-TS, ANSI and ETSI to begin the development of corresponding non-ISDN Frame Relay standards (X.36). These are basically the same as their ISDN counterparts, but with ISDN dependencies removed.

The earliest public Frame Relay implementations offer only PVC services, in effect providing a virtual circuit alternative to leased lines but at significantly lower costs. The greater economies of scale that public network operators achieve mean that they can usually offer higher levels of service quality (shorter delays, lower frame loss rates, etc.) than an equivalent private network. Where a private network would typically use T1/E1 links between switching nodes, a public network operator would be able to use T3/E3 to achieve substantial performance advantages.

Largely for regulatory reasons, public Frame Relay services are generally available in two forms, often referred to as bundled and unbundled, as shown in Figure 3.9.

In the bundled service the network operator provides the terminating routers as well as the PVCs. The user interface to the service is on the LAN side of the router so that a true LAN interconnection service is provided, e.g. ethernet-to-ethernet, token ring-to-token ring, etc. Clearly there is scope for providing added-value services such as ethernet-to-token ring interworking. In the unbundled service the user provides and maintains the routers and the service provided is a 'native-mode' Frame Relay service. To ensure interoperability the network operators generally insist on the use of known, 'certificated', router equipment.

It is important to keep in mind that substantial differences can exist between apparently similar Frame Relay services offered by different service

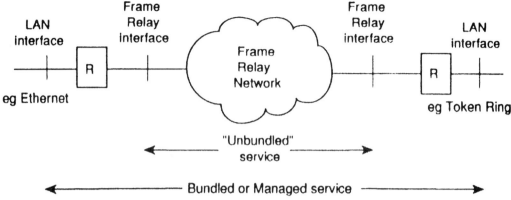

Figure 3.9 Bundled and unbundled Frame Relay services

providers (depending on interpretation of CIR, the terms of the 'guarantee', and so on), and the fine print of the service level agreement should be understood clearly by the potential user before committing.

LAN-interconnection using the Frame Relay data transmission service

Figure 3.10a shows a typical arrangement of LAN routers interconnected by leased lines to form a (very small) corporate data network. Full mesh interconnection of the routers is generally not cost-effective using leased lines so some of the routers carry 'transit' traffic (in this case traffic between B and C would be routed via A). Note that each Router needs as many wide area interfaces as there are leased lines connected to it. Note also that adding a new leased line to expand the network requires a visit from a technician to install and commission it.

Figure 3.10b shows the equivalent arrangement using Frame Relay PVCs instead of leased lines. By appropriate choice of service parameters (access rate, CIR, B_c, and B_e) the capacity of the PVCs can be tailored to closely match the actual traffic flows. However, there are two important differences, both favouring Frame Relay. Firstly, since one access circuit carries multiple PVCs, only one wide area interface is needed for each router, thus reducing router costs. It also becomes more attractive to move towards full mesh interconnection, achieving further savings in router costs because of the reductions in transit traffic that the routers need to be dimensioned to carry.

Secondly, only the access circuit between the router and the serving FR switch is dedicated to the user. Circuits between the FR switches are statistically shared with many other users, making substantial reductions in transmission costs possible. Furthermore, the access circuits will get shorter as public Frame Relay networks grow and more 'points of presence' are installed, so this cost advantage will continue to get better.

What is more, these cost savings need not bring a performance penalty. Since the frames are everywhere transmitted at the full rate of the bearer circuit the cross-network delays can be kept short. In general the bearer circuits are of higher speed than the leased line alternatives would be. In contrast with leased lines, additional PVCs can be added using remote

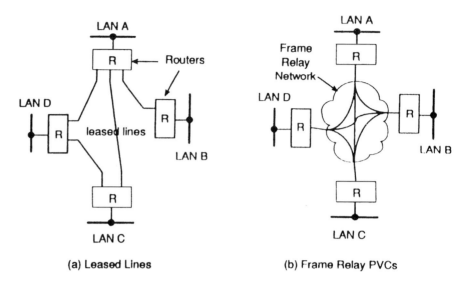

Figure 3.10   LAN Interconnection using the Frame Relay data transmission service

management procedures. No changes to existing hardware are generally needed. Also, as the pattern of traffic around the corporate network changes, reflecting the evolution of the company, the throughput and configurations of the PVCs can quickly and easily be modified to match using remote management procedures.

Real corporate data networks of course are very much larger than this simple example, and may involve hundreds of sites and thousands of terminals. As they continue to grow in size, speed and complexity they become increasingly difficult to manage and maintain, placing ever greater demands on the routers, both in terms of features and throughput, and on the network management systems. By facilitating greater mesh interconnection and more flexible centralised management, Frame Relay can help to contain and reduce this complexity.

By buying managed Frame Relay LAN interconnection services from public network operators, much of the complexity and many of the problems can be handed over to someone else, leaving the company to focus on its core business. The late 1980s saw corporate telephone traffic begin to migrate from private networks on to virtual private networks based on public switched networks. It looks as though Frame Relay is now set to do the same thing for corporate data networks.

Carrying multiple LAN protocols over Frame Relay

We have seen that one of the great merits of the simple Frame Relay data transfer protocol is its transparency to higher-layer protocols. A LAN packet is simply placed into the user information field of a Frame Relay frame at the sending end and taken out unchanged at the receiving end; the Frame Relay network neither knows nor cares what is in the user information field. This

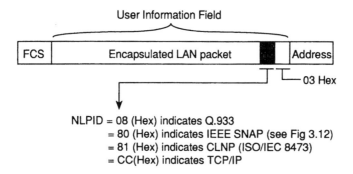

Figure 3.11   Higher-layer protocol identification using NLPIDs

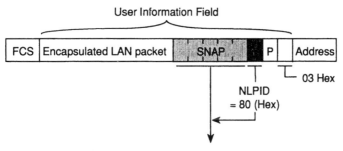

Figure 3.12   Higher-layer protocol identification using SNAP identifiers

means that multiple LAN protocols (e.g. TCP/IP, XNS, IPX, DECnet, AppleTalk, OSI, etc.) can be carried over a single virtual circuit, whether PVC or SVC.

But the receiving end needs to know what LAN protocol is being carried in each frame so that the right processing can be carried out; different LAN protocols need different treatment. Clearly some indication needs to be carried with each frame to identify the higher-layer protocol being carried. A standard has been created for doing this (T1.617, FRF.3) in which a protocol identifier is included at the beginning of the user information field.

ISO and ITU-TS administer a scheme of network level protocol identifiers (NLPIDs) which identify standardised protocols used by LANs (ISO/IEC TR 9577). The NLPID is included as the second octet in the user information field of the Frame Relay frame as shown in Figure 3.11.

LAN protocols that do not fall within the scope of ISO and ITU-TS cannot be handled quite so simply because they have not been assigned specific NLPIDs. However, most of them, including the IEEE 802 standards and proprietary protocols such as IPX, AppleTalk, etc. have been assigned five-octet sub network access protocol identifiers, generally referred to as SNAPs. These protocols are identified by assigning a specific NLPID to indicate that a SNAP identifier follows, as shown in Figure 3.12.

Whatever form the intervening circuit takes, LAN interconnection is either done by bridges, which deal in MAC layer frames containing LLC addresses, or by routers, which deal in the LAN's network layer frames containing Network layer addresses. Since by definition NLPIDs apply to network layer protocols, bridged frames will always use the SNAP identifier method; but routed frames may use either. They will use the NLPID encapsulation if an NLPID has been allocated to the LAN network layer protocol, but the SNAP encapsulation otherwise.

Not surprisingly, in view of the wide variety of LAN protocols that may be encountered, the subtleties of this scheme are considerable. For example, if the LAN packet is too long to fit into the longest user information field supported by a particular Frame Relay service (and remember that the default maximum for this is only 260 octets) it will be necessary to segment the packet at the sending end, transfer it as a sequence of Frame Relay frames, and reassemble it at the receiving end. Segmentation and reassembly of this sort is also included as part of the multiprotocol encapsulation scheme. The inquisitive reader is referred to the source documents for a detailed treatment.

Frame Relay interworking

So far we have treated interconnected LANs as closed systems. In reality of course, this is not the case, and all corporate data networks need to provide for interworking with external terminals and hosts. For example, it is a common requirement for remote low-speed asynchronous terminals or X.25 packet-mode terminals to gain switched access to a host on a LAN. Similarly, LAN-based terminals may wish access a remote TCP/IP host on the Internet.

A Frame Relay-X.25 gateway would typically be used to provide this interworking, as shown in Figure 3.13. The gateway terminates the X.25 protocol on one side and the Frame Relay protocol on the other. The contents of the user information field of X.25 packets are passed to the Frame Relay side where they form the user information field of the Frame Relay frames. If necessary the gateway also performs segmentation and reassembly to overcome any differences in the maximum packet/frame sizes supported by the X.25 and Frame Relay networks.

Another approach borrowed from X.25 is to use Frame Relay assembler/ disassembler devices (FRADs) to enable a non-Frame-Relay terminal to access the Frame Relay network.[1] A FRAD encapsulates the terminal's 'native' protocol (such as SNA/SDLC, BSC, X.25 or low-speed asynchronous) in Frame Relay frames for transmission through the Frame Relay network. A complementary FRAD would be needed at the remote end to recover the encapsulated information. Unlike the Gateway, the FRAD does not do any protocol conversion.

FRADs would typically be used to support X.25 packet services, SNA, and low-speed asynchronous data terminals. Though shown in Figure 3.13 as

---

[1] The reader should be aware that the term FRAD is also interpreted to mean Frame Relay access device, in which case even native-mode Frame Relay routers may be regarded as FRADs.

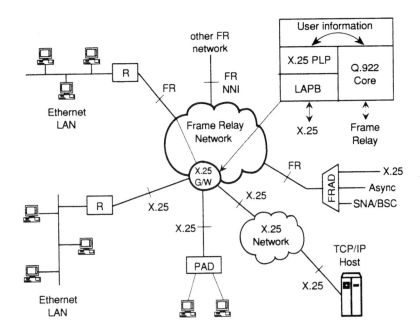

Figure 3.13 Interworking FR with other data services

located at the customer's premises, the FRAD may nevertheless be managed as part of the network to provide a managed service to the user. FRADs may also be deployed within the network, typically to support dial-up access to the Frame Relay services.

In practice FRADs are likely to include concentration capability, as shown, so that traffic from a group of terminals, including non-Frame-Relay terminals, can be cost-effectively delivered to a Frame Relay network.

It is also necessary for Frame Relay networks to interwork with other Frame Relay networks, and standards are being developed to facilitate this based closely on the user–network interface. The main differences relate to management capability which needs to be more comprehensive for the network–network interface.

But even with a standard network–network interface, interworking may not be entirely straightforward because of detailed differences that may exist in the way that different makes of Frame Relay switches work. For example, to provide an acceptable level of security against the possible failure of an internetwork link it may be necessary to have more than one point of interconnection. The problem then arises of how to arrange for the rapid automatic reconfiguration of existing virtual circuits spanning both networks to maintain continuity of service should any of the internetwork links fail. Detecting the failure may be straightforward, but the two networks would need to co-operate closely to reconfigure the affected virtual circuits, and this is not likely to be straightforward.

Managing PVCs

A simple standard has been agreed for signalling PVC status over the user–network interface so that users can create and maintain a comprehensive picture of the state of the PVCs on that interface (Q.933, T617). The idea is that the user periodically sends a STATUS ENQUIRY message into the network which responds with a STATUS report message.

Two types of status enquiry are sent periodically. A link integrity verification status enquiry is used to check that the access circuit is still functioning (strictly speaking it verifies that the link with DLCI = 0, which is the link used to carry the PVC status signalling, is still functioning). A full status enquiry is used to elicit a report from the network on the status of every PVC carried on the access circuit.

The user would send a status enquiry typically every 10 seconds. Most would be link integrity verification enquiries, but every so often (typically every sixth request) a full status report would be requested. The response to a full status enquiry would indicate newly added PVCs and, by omitting a previously reported PVC, PVCs that have been deleted. The procedure ensures that a PVC cannot be deleted and another (perhaps to a different remote end-point) added using the same DCLI without the user being notified of the change. The network may respond to any status enquiry with a full status report if any PVC status has changed or new PVC(s) has been added.

Additionally, the network may send a single PVC asynchronous status enquiry at any time to notify the user of a change of the status of any individual PVC on the access circuit. Such messages may be sent as and when required, not in response to a status enquiry message.

The network may optionally seek status reports from the user using the same protocol.

If any changes are detected by the user, either explicitly or by comparing incoming status information with stored tables, appropriate changes would be made in the routers' routing tables to avoid unavailable PVCs.

Because there is generally a delay between the time the network makes a new PVC available and the time the user is told about it in a status report message, there is the possibility that users may receive frames on PVCs they do not know they have got (i.e. unknown DLCIs)! It is up to the users to resolve this. Similarly the network may receive frames on PVCs it thinks are not available. This is left for the network to sort out.

Sequence numbers are used in the status enquiry and status responses to guard against the possible effects of errors or lost frames.

If the network detects a condition at a user–network interface that affects service, such as not receiving a STATUS ENQUIRY message within a timed period, typically 15 seconds, it should notify the remote end of each service-affected PVC either in a status response to a full status enquiry message or, for a single PVC, in an *ad hoc* asynchronous status report message.

Switched virtual circuits (SVCs)

LAN interconnection is not really a random interconnection requirement. The pattern of interconnection is generally fixed and PVCs can satisfy the need. But some applications need switched connections, i.e. SVCs. In their infinite wisdom CCITT, in their ISDN-based standards for Frame Relay, had the foresight to permit call control signalling to be carried either in the separate ISDN D-channel or in-channel using DLCI = 0. The latter option can be used to achieve SVC services in stand-alone, non-ISDN, data networks, and there seems little doubt that these will be implemented before ISDN versions appear.

The signalling protocol used in-channel is the same as in the ISDN D-channel for Frame Mode services, Q.933. To facilitate early introduction of SVC services at minimum cost the Frame Relay Forum has defined a considerably simplified version of the full Q.933 signalling standard, colloquially and aptly known as the skinny subset (FRF.4).

SVC capability will considerably expand the range of user applications that Frame Relay will be able to support. With the increasing penetration of error-free digital transmission and greater intelligence in terminal equipment it is likely that with SVCs Frame Relay will begin to displace X.25 as a bearer service, even at the lower data rates, though it is also likely that the X.25 packet layer protocol will continue to be used, operated on an end-to-end basis between the terminals.

3.4   SUMMARY

This chapter has introduced the main features of Frame Relay, a wide area data service with a very simple data transfer protocol designed for use at high data rates (up to 34 Mbit/s) and providing a high degree of transparency to higher layer protocols operating between the users' end systems. Though originally created as an ISDN bearer service it has rapidly developed into a data service in its own right and non-ISDN Frame Relay networks are now becoming widespread, both as private networks and as public networks. Initial implementations, being aimed at the LAN interconnection market, offer PVC services which give a very flexible and cost-effective alternative to leased lines, especially where three or more geographically dispersed sites need interconnection.

We have looked at the ISDN roots of Frame Relay, which explains why it has the protocol architecture it has, and we have outlined its evolution as a non-ISDN data transmission service. The lightweight data transfer protocol has been explained, together with how it is 'packaged' as a defined service with a policed quality of service. We have also seen something of how the user can keep track of what his PVCs are up to, or whether they are up to anything at all! In view of Frame Relay's importance for LAN interconnection we have also seen fit to describe how multiple LAN protocols can be carried over a virtual connection and treated appropriately at the receiving end.

REFERENCES

General

Griffiths, J. M. (1992) *ISDN Explained*, John Wiley & Sons.
Hopkins, H. H. (1993) *The Frame Relay Guide*. CommEd.

Standards

ITU-TS (née CCITT)

E.164 Numbering Plan for the ISDN Era
I.122 Framework for Frame Mode Bearer Services
I.233.1 ISDN Frame Relaying Bearer Service
I.233.2 ISDN Frame Switching Bearer Service
I.370 Congestion Management for the ISDN Frame Relaying Bearer Service
I.372 Frame Mode Bearer Service Network-to-Network Interface Requirements
I.555 Frame Mode Bearer Service Interworking
Q.921 ISDN User–Network Interface - Data Link Layer Specification
Q.922 ISDN Data Link Layer specification for Frame Mode Bearer Services
Q.931 ISDN User–Network Interface Layer 3 Specification for Basic Call Control
Q.933 DSS1—Signalling Specification for Frame Mode Basic Call Control
X.36   Interface between data terminal equipment (DTE) and data circuit-terminating equipment (DCE) for public data networks providing Frame Relay Data Transmission Service by dedicated circuit
X.76   Network-to-Network Interface between PDNs providing the Frame Relay Data Transmission Service

ANSI T1S1

T1.606 Frame Relay Bearer Service—Architectural Framework and Service Description
T1.606 Addendum: Congestion Management Principles
T1.617 DSS1—Signalling Specification for Frame Relay Bearer Service
T1.618 DSS1—Core Aspects of Frame Protocol for Use with Frame Relay Bearer Service

ETSI

ETS 300 399-1   Frame Relay Service—General Description
ETS 300 399-2   ISDN   Frame Relay Bearer Service—Service Definition
ETS 300 399-3   Frame Relay Data Transmission Service—Service Definition
ISO/IEC TR 9577   Protocol Identification in the Network Layer

Frame Relay Forum

FRF.1   Frame Relay User-to-Network Interface (UNI) Implementation Agree-
        ment
FRF.2   Frame Relay Network-to-Network Interface (NNI) Phase 1
        Implementation Agreement (deals with PVCs only)
FRF.3   Multiprotocol Encapsulation over Frame Relay Implementation
        Agreement
FRF.4   Frame Relay User-to-Network SVC Implementation Agreement

4

Switched Multi-megabit Data Service (SMDS)

My [foreign] policy is to be able to take a ticket at Victoria
Station and go anywhere I damn well please

Ernest Bevin

Many companies espouse a similar policy in relation to their data communications. They have paid dearly for their ticket and want unfettered and reliable routes for their information. Driven by the growing importance of data communications to business operations and the trend towards global operation, the demand for effective wide area data communications is increasing rapidly. Corporate data networking, mainly interconnecting the LANs located at a company's various sites, is a growing and valuable market and customers' requirements and expectations are growing with it. SMDS, the brainchild of Bellcore, the research arm of the Regional Bell Operating Companies (RBOCs) in the USA, has been developed to meet these needs.

SMDS, as the switched multi-megabit data service is invariably known, is a public, high-speed, packet data service aimed primarily at the wide area LAN interconnection market. It is designed to create the illusion that a company's network of LANs (its corporate LAN internet) is simply one large seamless LAN.

It has one foot in the LAN culture in that, like LANs, it offers a connectionless service (there is no notion of setting up a connection—users simply exchange information in self-contained variable-length packets), and there is a group addressing option analogous to the multicast capability inherent in LANs that enables a packet to be sent simultaneously to a group of recipients.

But, it also has a foot firmly in the public telecommunications culture in that it uses the standard ISDN global numbering scheme, E.164, and it provides a well-defined service with point-of-entry policing—the SMDS access path is dedicated to a single customer and nobody can have access to anybody else's information—so that, even though based on a public network, SMDS offers the security of a private network.

Because it is a connectionless service every packet contains full address information identifying the sender and the intended recipient (or recipients in the case of group addressing). Because it is policed at the point-of-entry to the network these addresses can be screened to restrict communication to selected users to provide a Virtual Private Network (VPN) capability.

In effect SMDS offers LAN-like performance and features over the wide area, potentially globally. It is therefore an important step on the road to Total Area Networking, eroding as it does the spurious (to the user) distinction between local and wide area communications.

SMDS specifications are heavy, both physically and intellectually! The aim here is to provide an easy introduction for those who want to know what SMDS is but who are not too bothered about how it's done, and at the same time to satisfy readers who want to dig a bit deeper. The description is therefore covered in two passes. In section 4.1 we outline the main features of SMDS, enough to give a basic understanding. In section 4.2 we provide more detail for those who want it. In section 4.3 we cover a few miscellaneous issues that help to tie it all together.

4.1   THE BASICS OF SMDS

There is always jargon! The SMDS access interface is called the Subscriber Network Interface or SNI, and the protocol operating over the SNI is called the SMDS Interface Protocol or SIP, as shown in Figure 4.1, which illustrates a typical example of how SMDS is used. It shows SMDS interconnecting a LAN with a stand-alone host. But from this the reader can mentally visualise interworking between any combination of LAN–LAN, LAN–host or host–host.

The SMDS service interface is actually buried in the CPE. In the example in Figure 4.1 it is the interface between the customer's internet protocol and the SIP. To provide a high degree of future-proofing the SMDS service is intended to be technology independent. The idea is that the network technology can be upgraded (for example, to ATM) without making the SMDS service obsolete. The implementation of the SMDS Interface Protocol would have to be upgraded to track changes in network technology: the purpose of the SIP is after all to map the SMDS service on to whatever network technology is being used. But, because the SMDS service seen at the top of the SIP remains the same, an SMDS customer using one network technology would still be able to interwork fully with a customer using a different technology.

SMDS packets—the currency of exchange

The SMDS service supports the exchange of data packets containing up to

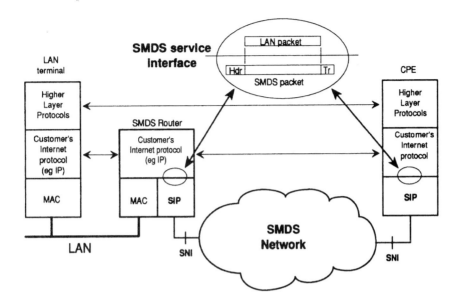

Figure 4.1 The SMDS service

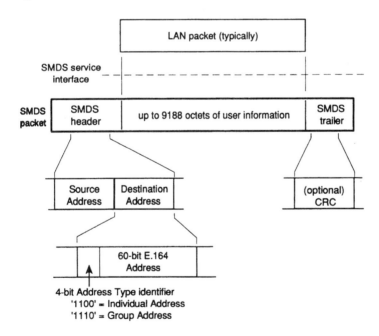

Figure 4.2 The SMDS packet

9188 octets of user information, as shown in Figure 4.2. This user information field would typically contain a LAN packet. Why 9188 octets? Because it can accommodate just about every type of LAN packet there is.

The header of the SMDS packet contains a source address identifying the sender and a destination address identifying the intended recipient (it also contains some other fields that we will look at in section 4.2). The destination address may be an individual address identifying a specific Subscriber Network Interface (SNI), or it may be a group address, in which case the SMDS network will deliver copies of the packet to a pre-agreed list of remote network interfaces.

The SMDS packet trailer contains (amongst other things also covered in section 4.2) an optional four-octet CRC field enabling the recipient to check for transmission errors in the received packet. This CRC may be omitted; if error checking is done by a higher layer protocol, for example, it would clearly be more efficient, and faster, not to do it again in the SMDS layer.

SMDS addressing

An SMDS address is unique to a particular SNI, though each SNI can have more than one SMDS address. Typically each piece of CPE in a multiple-CPE arrangement (see below) would be allocated its own SMDS address so that SMDS packets coming from the network can be picked up by the appropriate CPE (though strictly speaking it is entirely the customer's business how he uses multiple-SMDS addresses that are assigned to him).

In the case of group addressing, the destination address field in delivered SMDS packets will contain the original group address, not the individual address of the recipient. The recipient then knows who else has received this data, which may be important in financial and commercial transactions. The network will only deliver a single copy of an SMDS packet to a particular SNI, even if the group address actually specifies more than one SMDS address allocated to that SNI. It is for the CPE to decide whether it should pick up an incoming SMDS packet by looking at the group address in the destination address field.

The destination and source address fields in the SMDS packet actually consist of two sub-fields, as shown in Figure 4.2, a 4-bit address type identifier and a 60-bit E.164 address. The E.164 numbering scheme does not directly support group addressing, so the address type identifier is needed to indicate whether the associated address field contains an individual address or a group address. Since group addressing can only apply to the destination address, the address type identifier in the source address field always indicates an individual address.

The E.164 number, which may be up to 15 decimal digits long, consists of a country code, which may be of 1, 2 or 3 digits, and a national number of up to 14 digits. The structure of the national number will reflect the numbering plan of the country concerned

A group address identifies a group of individual addresses (typically up to 128), and the network will deliver group addressed SMDS packets to each SNI that is identified by any of the individual addresses represented by the group address, except where any of these individual addresses are assigned

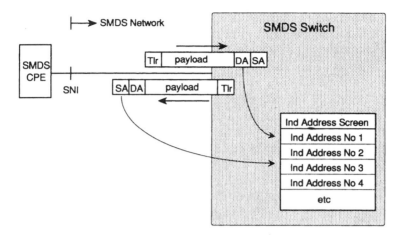

Figure 4.3   Address screening

to the SNI which sent the SMDS packet. The SMDS network will not send the SMDS packet back across the SNI which sent it.

A particular individual address may be identified by a number of group addresses (typically up to 32), so that a user may be part of more than one virtual private network or work group.

Address screening—building virtual private networks

Because every Subscriber Network Interface is dedicated to an individual customer, the network is able to check that the source addresses contained in SMDS packets sent into the network from a particular SNI are legitimately assigned to that customer. If this check fails the network does not deliver the packet. This point-of-entry policing prevents the sender of an SMDS packet from indicating a fraudulent source address and the recipient can be sure that any SMDS packet he receives is from the source address indicated.

SMDS also provides a facility for screening addresses to restrict delivery of packets to particular destinations. For this purpose the network uses two types of address lists: individual address screens, and group address screens. An address screen relates to a specific Subscriber Network Interface and is agreed with the customer at subscription time.

An individual address screen, which can contain only individual addresses, is used for screening the destination addresses of SMDS packets sent by the CPE and the source addresses of packets to be delivered to the CPE, as shown in Figure 4.3. Individual address screens contain either a set of 'allowed' addresses or a set of 'disallowed' addresses, but not both. When the screen contains allowed addresses, the packet is delivered only if the screened address matches an address contained in the address screen. Similarly, when the screen contains disallowed addresses the packet is delivered only if the screened address does not match an address contained in the address screen.

The group address screen, which can contain only group addresses, is used to screen destination addresses sent by the CPE. It also contains either a set of 'allowed' adresses or a set of 'disallowed' addresses, and is used in a similar way to the individual address screen as described above.

In some implementations the network may support more than one individual or group address screen per subscriber network interface. But if more than one individual or group address screen applies to an SNI the customer must specify which address screens are to be used with which of the SNI's addresses.

Use of address sceening enables a company to exercise flexible and comprehensive control over a corporate LAN internet implemented using the SMDS and be assured of a very high degree of privacy for its information. In effect a customer can reap the cost-performance benefits arising from the economies of scale that only large public network operators can achieve, while having the privacy and control normally associated with private corporate networks.

Tailoring the SMDS to the customer's needs—access classes

Customers will have a wide variety of requirements both in terms of the access data rates (and the corresponding performance levels) they are prepared to pay for and the traffic they will generate. Four access data rates have so far been agreed for the SMDS, DS1 (1.544 Mbit/s) and DS3 (44.736 Mbit/s) for use in North America and E1 (2.048 Mbit/s) and E3 (34 Mbit/s) for use in Europe. Higher rates are planned for the future including 140/155 Mbit/s.

For the higher access data rates, that is 34 and 45 Mbit/s, a number of access classes have been defined to support different traffic levels, as summarised below. This arrangement enables the customer to buy the service that best matches his needs; and it enables the network operator to dimension the network and allocate network resources cost-effectively. SIR stands for sustained information rate and is the long-term average rate at which information can be sent.

Access Class	SIR (Mbit/s)
1	4
2	10
3	16
4	25
5	34

Note that Access Classes 1–3 would support traffic originating from a 4 Mbit/s Token Ring LAN, a 10 Mbit/s Ethernet LAN, and a 16 Mbit/s Token Ring LAN, respectively (though the access classes do not have to be used in this way, it is really up to the customer).

The network enforces the access class subscribed to by throttling the flow of SMDS packets sent into the network over an SNI. This uses a credit manager mechanism as described in section 4.2. There is no restriction on the traffic flowing from the network into the customer; and there is no access class enforcement for the lower access data rates, DS1 and E1, where a single access class applies. Access classes 4 and 5 also do not actually involve any rate enforcement since 25 Mbit/s and 34 Mbit/s are the most that can be achieved respectively, over E3 and DS3 access links because of the overheads inherent in the SMDS Interface Protocol.

The SMDS interface protocol

It can be seen from Figure 4.1 that the SIP is equivalent to a MAC protocol in terms of the service it offers to higher-layer protocols. To keep development times to a minimum and avoid the proliferation of new protocols, it was decided to base the SIP on a MAC protocol that had already been developed. It is in fact based on the standard developed by the IEEE for Metropolitan Area Networks (MANs) (ISO/IEC 8802-6).

In effect a MAN is a very large LAN. It is a shared-medium network based on a duplicated bus (one bus for each direction of transmission) and uses a medium access control (MAC) protocol known as Distributed Queue Dual Bus (DQDB), which can achieve a geographical coverage much greater than that of a LAN. DQDB provides for communication between end systems attached to the dual bus (we will call them nodes) and supports a range of services including connectionless and connection-oriented packet transfer for data and constant bit-rate (strictly speaking isochronous) transfer for applications such as voice or video. The SMDS interface protocol, however, supports only the connectionless packet transfer part of the DQDB capability. The DQDB protocol operating across the subscriber network interface is usually called the 'access DQDB'.

A DQDB network can be configured either as a looped bus (which has some self-healing capability if the bus is broken) or as an open bus arrangement. The access DQDB used in SMDS is the open bus arrangement. As shown in Figure 4.4, the access DQDB can be a simple affair, supporting a single piece of CPE; or it can support multiple-CPE configurations. In order to provide point-of-entry policing an SMDS access is always dedicated to a single customer, and all CPE in a multiple-CPE configuration must belong to that customer.

In the case of multiple-CPE configurations the access DQDB can support direct local communication between the pieces of CPE without reference to the SMDS network. So the access DQDB may be simultaneously supporting direct communications between the local CPE and communication between the CPE and the SMDS switch to access to the SMDS service. CPE that conforms to the IEEE MAN standard can be attached to the access DQDB and should in principle be able to use the SMDS service without modification.

The SMDS Interface Protocol is layered into three distinct protocol levels, as

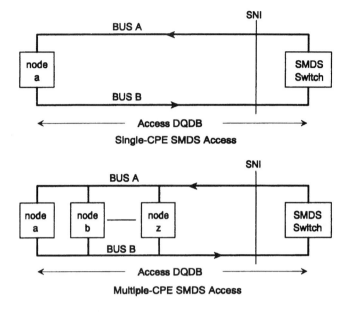

Figure 4.4   Access DQDB

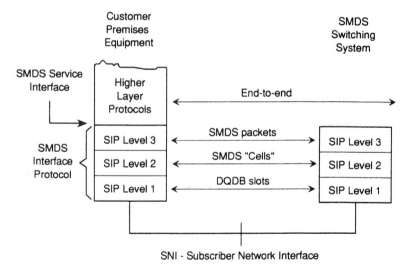

Figure 4.5   SMDS Interface Protocol (SIP)

shown in Figure 4.5. Though based on similar reasoning, these protocol levels do not correspond to the layers of the OSI reference model. As indicated in Figure 4.1, the three levels of the SIP together correspond to the MAC sublayer, which in conjunction with the LLC sublayer corresponds to the OSI Link Layer.

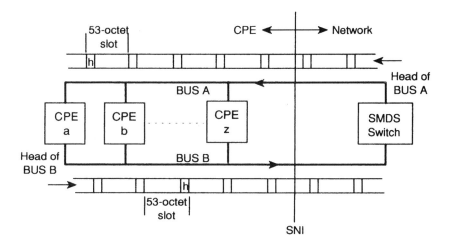

Figure 4.6 Slot structure on the DQDB bus

SIP Level 3

SIP Level 3 takes a data unit from the SMDS user, typically a LAN packet, adds the SMDS header and SMDS trailer to form an SMDS packet of the format shown in Figure 4.2, and passes it to SIP level 2 for transmission over the SNI. In the receive direction Level 3 takes the SMDS packet from Level 2, performs a number of checks on the packet, and if everything is OK passes the payload of the packet (that is the user information) to the SMDS user.

SIP Level 2

SIP Level 2 is concerned with getting the Level 3 SMDS packets from the CPE to the serving SMDS switch, and vice versa, and operates the DQDB access protocol to ensure that all CPE in a multiple-CPE configuration get a fair share of the bus.

On each bus DQDB employs a framing structure of fixed-length slots of 53 octets, as shown in Figure 4.6. This framing structure enables a number of Level 3 packets to be multiplexed on each bus by interleaving them on a slot basis. In this way a particular piece of CPE, or a multiple-CPE configuration, can send a number of SMDS packets concurrently to different destinations (and receive them concurrently from different sources). The slot structure means that a node does not have to wait until a complete Level 3 packet has been transferred over the bus before the next one can start to be transmitted.

We will defer description of the DQDB access control protocol to section 4.2. Even a simplified account of this would be a bit heavy going for the casual reader. It is enough here to know that the DQDB access control protocol gives each piece of CPE in a multiple-CPE arrangement fair access to the slots on the buses to transfer packets, either to another piece of local CPE or, if using the SMDS service, to the SMDS switch.

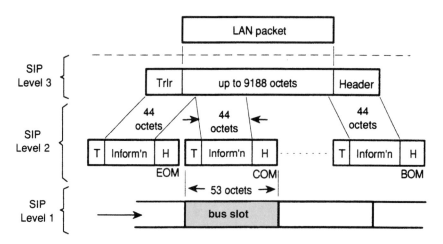

Fig 4.7 Segmentation and reassembly

For transmission over the access DQDB, SIP Level 2 has to chop the SMDS packets up to fit into the 53-octet DQDB slots, a process usually referred to as segmentation. In the receive direction, it reassembles the received segments to recreate the packet to pass on to SIP level 3.

The segmentation and reassembly process operated by SIP level 2 is outlined in Figure 4.7. The SMDS packet passed from SIP level 3 is divided into 44-octet chunks, and a header of 7 octets and trailer of 2 octets is added to each to form a 53-octet Level 2 data unit for transport over the bus in one of the slots. If the SMDS packet is not a multiple of 44 octets (and usually it will not be) the last segmentation unit is padded out so that it also consists of 44 octets. The information in the level 2 header and trailer are used at the remote end to help reassemble the SMDS packet. Reassembly is the converse of the segmentation process. We will take a more detailed look at segmentation and reassembly in section 4.2.

SIP level 1

SIP level 1 is concerned with the physical transport of 53-octet level 2 data units in slots over the bus. For our purposes we will assume that each bus carries contiguous slots, though strictly speaking the framing structure also includes octet-oriented management information interleaved with the 53-octet slots.

4.2 COMPLETING THE PICTURE

The above outline of SMDS will satisfy the needs of some readers, but others will want more detail. They should read on.

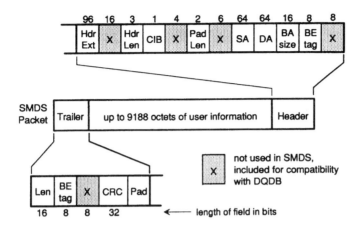

Figure 4.8   SMDS packet detail

The SMDS packet in detail

Figure 4.8 shows the various fields in the SMDS packet. The numbers indicate how many bits the fields contain. Fields marked X are not used in SMDS and are included to maintain compatibility with the DQDB packet format: the SMDS network ignores them. The SMDS service delivers the complete SMDS packet to the recipient as sent. It should make no changes to any of the fields.
   The purpose of each field is described briefly below.

- BEtag: the 1-octet BEtag (Beginning End tag) fields have the same value in both the header and trailer. They are used by the recipient to check the integrity of the packet.

- BASize: this 2-octet field indicates the length of the SMDS packet, in octets, from the destination address field up to and including the CRC field (if present); it is used by the credit manager mechanism used to enforce the access class.

- Destination address: this contains the E.164 address of the intended recipient, either as an individual address or as a group address. The detailed format is shown in Figure 4.2.

- Source address: this contains the E.164 address of the sender.
  (Non-SMDS DQDB packets normally transport MAC-layer packets in which case the destination and source address fields contain MAC addresses.)

- Pad Len: this 2-bit field indicates the number of octets of padding (between 0 and 3) inserted in the trailer to make the complete packet a multiple of 32 bits (for ease of processing).

- CIB: this 1-bit field indicates whether the optional CRC field is present in the trailer (CIB = 1), or not (CIB = 0).

- Hdr Len: this 3-bit header extension length field indicates the length of the header extension field (it actually counts the header extension in multiples of 32 bits): in SMDS the header extension is always 12 octets, so this field always contains binary 011).

- Hdr Ext: this 12-octet header extension field is used in some implementations of SMDS to indicate the SMDS version number being used (recognising the fact that as SMDS evolves to provide enhanced features different versions will need to interwork), and to indicate the customer's choice of network carrier; if these features are not implemented the complete field is set to 0.

- PAD: this field may be from 0 to 3 octets long, and is used to make the complete SMDS packet a multiple of 32 bits (for ease of processing).

- CRC: if present (see CIB field) this field contains a 32-bit error check covering the packet from the destination address to the end of the packet.

- Len: this 2-octet field is set equal to the BAsize and is used by the recipient to check for errors in the packet.

The receiving level 3 process, which may be in an SMDS switch or the recipient's CPE, checks the format of the received packets. If any of the fields are too long or too short, or have the wrong format, or contain values that are invalid (such as header extension length field not equal to 011, or a BAsize value greater than the longest valid packet) the associated packet is not delivered (if an error is found by the SMDS switch) or should not be accepted (if an error is found by receiving CPE).

The credit manager

Customers with the higher speed (DS3 or E3) access circuits may subscribe to a range of access classes as outlined in section 4.1. The Access Class determines the level of traffic the customer may send into the SMDS network (the level of traffic received is not constrained in this way). The permitted traffic level is defined by a credit manager mechanism operated by the serving SMDS switch. If the traffic level subscribed to is exceeded, the network will not attempt to deliver the excess traffic.

To avoid packet loss it may be desirable for CPE to implement the same credit manager algorithm so that the agreed access class is not violated (however it should be noted that, since the access class relates to the SNI and not to CPE, this option is really only available in single-CPE configurations). But this is entirely the customer's business: it is not a requirement of SMDS.

The credit manager operates in the following way.

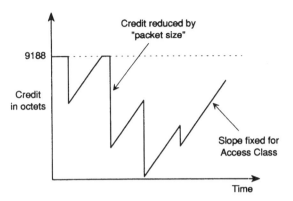

Figure 4.9 The SMDS credit manager

A customer is given 9188 octets of 'credit' to start with, enough to send the longest possible SMDS packet. When he sends a packet into the SMDS network he uses credit equal to the number of octets of user information contained in the packet, reducing his credit balance. At the same time the customer continuously accrues credit at a rate determined by the access class he has subscribed to. The higher the access class the faster credit is accrued.

When he then sends another packet the serving SMDS switch checks whether he has got enough credit to cover it; that is, has he got as many octets of credit as there are octets of user information in the packet? If he has enough credit the packet is accepted for delivery and his credit balance is reduced by however many octets of user information it contains. If he has not got enough credit the packet is not accepted for delivery and his credit is not debited. And so on for successive packets.

For ease of implementation (BASize-36) is actually used as a close approximation to the number of octets of user information contained in a packet. So, having illustrated the principle, the reader should now substitute (BASize-36) for the number of octets of user information in the above description.

The credit manager is illustrated graphically in Figure 4.9. This shows that the maximum possible credit is 9188 octets. And it shows that credit is continuously accrued at a rate fixed by the access class in force. This is a bit of a simplification. Credit is actually accrued at discrete, but frequent, intervals and is incremented in discrete steps.

The information needed by the SMDS switch to operate the credit manager is contained in the first level 2 segment carrying a packet. The switch does not have to wait until the complete packet has been received. If a problem with a subsequent level 2 segment carrying part of the packet (such as a format or transmission error) means that the packet cannot be delivered, the credit subtracted for this packet is not returned.

Furthermore, the credit manager operates before any address validation or screening is done, and if these result in a packet not being delivered the credit subtracted for the packet is, likewise, not returned.

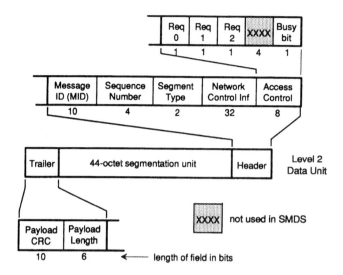

Figure 4.10   Level 2 Data Unit

Segmentation and reassembly

We have seen in section 4.1 that SMDS packets are segmented by the CPE for transfer over the DQDB-based SNI to the serving SMDS switch, and in the other direction of transmission are reassembled from the incoming segments. To understand how this is achieved we need to look into the header and trailer of the 53-octet Level 2 data unit shown in detail in Figure 4.10.

The 8-bit access control field is used by the DQDB access control protocol as described in the next section. It is not really involved in the segmentation and reassembly process. Nor is the network control information field which is included to maintain compatibility with DQDB (it is used to support the DQDB isochronous service). The purpose of the other fields is as follows.

- Segment Type: this field tells the receiving level 2 process how to process the segment (strictly speaking the term segment refers to the least significant 52 octets of the level 2 data unit shown in Figure 4.10; it does not include the access control field).

  A segment may be of four types. SMDS packets up to 44 octets long can be transferred in a single segment. This is identified as a single-segment message (SSM). Packets up to 88 octets long will fit into two segments. The first of these is identified as a beginning of message (BOM) segment, the second as an end of message (EOM) segment. Packets longer than 88 octets require additional intermediate segments between the BOM and EOM segments: these are identified as continuation of message (COM) segments.

- Sequence number: for any SMDS packet that needs more than one segment the 4-bit sequence number identifies which segment it is in the sequence. The sequence number is incremented by one in each successive segment carrying a particular packet, and enables the receiving process to tell whether any segments have been lost.

- Message identifier (MID): the value in this 10-bit field identifies which packet in transit over the SNI the associated segment belongs to. It is the same in all segments carrying a particular packet.

  For a single-CPE configuration, administration of the MID is simple. To send a new packet to the CPE, the SMDS switch would select a currently unused MID value in the range 1 to 511. Before sending a packet into the SMDS network, the CPE would choose from the range 512 to 1023.

  For a multiple-CPE configuration the situation is not so straightforward because a piece of CPE must not use a MID value that is already in use, or currently being prepared for use, by another piece of CPE. In this case the allocation of MID values to different pieces of CPE is administered by a DQDB management procedure known as the MID page allocation scheme. It will not be described here, but it uses management information octets that are interleaved on the bus between the 53-octet slots carrying information. The key point is that all packets concurrently in transit over the SNI must have different MID values.

- Payload length: SMDS packets in general are not multiples of 44 octets. So SSM and EOM segments are likely to carry less than 44 octets of real information, the balance will be padding. The 6-bit payload length field identifies how much of the 44-octet segmentation unit is real information, and it will be used by the receiving process to identify and discard the padding. In BOM and COM segments the payload length field should always indicate 44 octets of real information.

- Payload CRC: this field contains a 10-bit CRC for detecting transmission errors. It covers the segment type, sequence number, MID, segmentation unit, payload length and payload CRC fields. It does not cover the access control or network control information fields.

To tie all this together it is instructive to put ourselves in the position of a piece of CPE in a multiple-CPE configuration (we will ignore local DQDB communication between the CPE and consider only SMDS traffic). We see a continuous sequence of slots passing on Bus A sent by the SMDS switch (see Figure 4.6); some are empty, some are full (as we will see in the next section, if the Busy bit = 0 the slot is empty, if the Busy bit = 1 the slot is filled). We take a copy of the contents of all filled slots. We are now in a position to answer the following questions.

Q If we are currently receiving more than one packet from the SMDS switch, how do we know which segments belong to which packets?

A The answer lies in the MID field; all segments belonging to the same packet will carry the same MID value.

Q How do we know when we have received all the segments carrying a particular packet? The answer here lies partly in the segment type field; if the packet is not carried in an SSM segment then we know we have received the complete packet when we have received the EOM segment with the MID value associated with that packet.

A It lies partly in the sequence number which tells us whether any segments have been lost in transit (there are also checks that the level 3 process can do on the 'completed' packet).

But there is another question we need to consider:

Q How do we know whether BOM segments and SSM segments passing on the bus are intended for us?

A The answer is that we have to look at the destination address field of the packet (remember that the complete SMDS packet header is contained in the first segment, that is the BOM segment, of a multiple-segment packet, or in the SSM segment if a single-segment packet). So we have to look not only at the header and trailer of the segments: if they are BOMs or SSMs we also have to look at the segmentation unit carried in the segment to find the destination address.

Having established that a particular packet is intended for us, we can of course identify subsequent COM or EOM segments belonging to that packet, because they will contain the same MID value as the BOM.

Receiving segments is straightforward: copying the information as it passes on the bus is a purely passive affair and does not need any access control. But sending segments needs to be carefully orchestrated. Access control is needed to prevent more than one node on the bus trying to fill the same slot and to make sure that all nodes get fair access.

DQDB access control protocol

DQDB uses two buses, shown as A and B in Figure 4.6, which each provide one direction of transmission. The two buses operate independently and together they support two-way communication between any pair of nodes. Data on each bus is formatted into contiguous fixed length slots, each of 53 octets (management information formatted in octets is actually interleaved with the slots to pass management information between the nodes, for simplicity we ignore this here). These slots carry the level 2 segments. The bus slot structure is generated by the nodes at the head of each bus. Data flow stops at the end of each bus (but not necessarily the management octets, some of which are passed on to head of the other bus).

In the SMDS application the switch is the node at the head of bus A. The other nodes are CPE, the end one of which performs as head of bus B. The SMDS subset of the DQDB protocol is used to transport SMDS packets between the CPE and the switch, as a sequence of segments carried in the DQDB slots, and from the switch to the CPE. But in the multiple-CPE configuration the DQDB protocol also permits communication between any pair of CPE, and this should co-exist with the SMDS service without

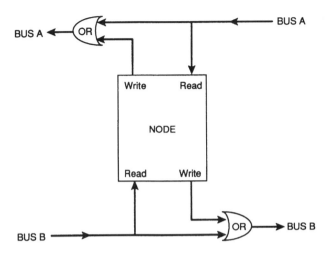

Figure 4.11 Bus access

interference. So the multiple-CPE arrangement can provide both local communications between the CPE and access to the SMDS service.

In order to communicate with another local CPE unit a piece of CPE needs to know where that other CPE unit is on the bus in relation to itself. For example, in Figure 4.6, if CPE b wanted to communicate with CPE a it would send information on bus A simply because CPE a is upstream on bus B. Similarly, if CPE b wanted to communicate with CPE z it would have to send the information on bus B. If the buses are not heavily loaded, the problem can be avoided by sending the same information on both buses. But if this is not acceptable then each piece of participating CPE will need to maintain a table of where the other nodes are in relation to itself. There are various ways of doing this, none particularly straightforward: we will duck the issue here and assume that these tables already exist.

For the SMDS use of the dual bus this issue does not arise since, from Figure 4.6, the switch, as the head of bus A, is always upstream of the CPE on bus A and is always downstream on bus B. So CPE always sends SMDS packets to the SMDS network on bus B and receives SMDS packets from the SMDS network on bus A.

Each node reads and writes to the bus as shown in Figure 4.11. A node can freely copy data from the buses but it cannot remove data from them. It may only put data on to a bus when the DQDB access control protocol allows it to. By implication this would only be when an empty slot is passing. The access control protocol operates independently for each bus. For clarity in the following explanation we will focus on how nodes send segments on Bus A.

The DQDB access control protocol uses the access control field in the header of the slot (see Figure 4.10). The busy bit is used to indicate whether the associated slot is empty (busy bit=0) or contains information (busy bit=1). The four bits marked XXXX are not used in SMDS. The three request bits, Req 0, Req 1 and Req 2, are used to signal that a node has information that it wants to send. Each request bit relates to a different level of priority. (In the

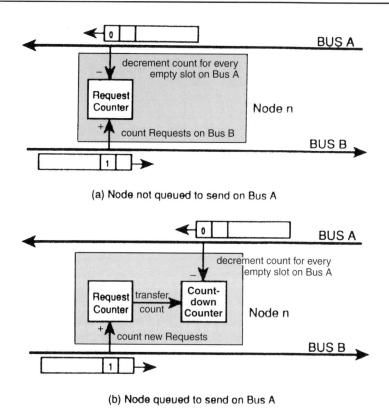

(a) Node not queued to send on Bus A

(b) Node queued to send on Bus A

Figure 4.12 DQDB access control

following description we will assume that all information queued for transmission on the buses has the same priority. This does not lose any of the story because the procedure described below is applied independently to each priority level).

It would clearly be grossly unfair if a node could simply fill the next empty slot to come along, since it would give that node priority over all other nodes further downstream. The aim of the DQDB access control protocol is to grant access to the bus on a first-come-first-served basis.

To achieve this each node needs to keep a running total of how many downstream nodes have requested bus access and not yet been granted it. Then, when a node wishes to gain bus access it knows how many empty slots it must let go past to service those outstanding downstream requests before it may itself write to the bus. The only way a node can know about requests made by downstream nodes is for those requests to be signalled on the other bus. So, in the DQDB access control protocol requests for access to bus A are made on bus B (and vice versa).

Each node uses a request counter as shown in Figure 4.12. This counter is incremented every time a slot goes past on bus B with the request bit set; this indicates a new request from a downstream node. It is decremented every time it sees an empty slot go past on bus A, this slot will service one of the outstanding downstream requests. So at any time the request counter indicates the number of unsatisfied downstream requests.

When a node has a segment ready for transmission on bus A it lets the upstream nodes know by setting the request bit in the next slot passing on bus B that does not already have the request bit set. At the same time it transfers the contents of the request counter to a countdown counter and resets the request counter. So at this point the countdown counter indicates the number of previously registered downstream requests that remain to be satisfied and the Request Counter begins to count new requests from downstream nodes.

The node may not write its own segment to the bus until all previous outstanding downstream requests (as indicated by the countdown counter) have been satisfied. So the countdown counter is decremented every time an empty slot goes past on bus A, and when the countdown counter reaches zero the node may write into the next empty slot that passes on bus A.

Using this procedure, operated by all nodes, the DQDB access control protocol gives every node fair access to the bus, closely approximating a first-come-first-served algorithm. We have illustrated how a node gets write access to bus A. The access control protocol is operated independently for each bus, and the nodes have a request counter and countdown counter for each direction of transmission.

SMDS performance and quality of service

To make meaningful comparisons with alternatives such as private networks, it is useful for the reader to have some idea of the levels of performance and service quality that SMDS aims to achieve. The objectives for these relate to availability, accuracy, and delay as follows. Note that they refer to the performance of a single network. In practice more than one network may be involved in delivering the service, particularly if international coverage is involved. So the figures should be regarded as the best that might be achieveable.

Availability

If you cannot use the service it is unavailable. Availability is the converse of unavailability! The objective is to achieve an availability of 99.9% for service on an SNI-to-SNI basis. This corresponds to a downtime of nearly 9 hours a year. Furthermore, there should not be more than 2.5 service outages per year, and the mean time to restore service should be no more than 3.5 hours.

Accuracy

Accuracy is about how often errors are introduced in the information being transported, assuming that the service is available! Performance objectives have been specified for errored packets, misdelivered packets, packets that are not delivered, duplicated packets, and mis-sequenced packets.

- *Errored*

A packet that is delivered with corrupted user information is an errored packet. The target is that fewer than 5 packets in 10^{13} will be errored. This assumes 9188 octets of user information per packet. It does not include undelivered packets or packets that are discarded by the CPE as a result of errors that can be detected by the SMDS interface protocol.

- *Misdelivered*

A packet is misdelivered if it is sent to the wrong Subscriber Network Interface. Fewer than 5 packets in 10^8 should be misdelivered.

- *Not delivered*

The SMDS network should lose fewer than 1 packet in 10^4.

- *Duplicated*

Only one copy of each packet should be delivered over an SNI (not including retransmissions initiated by the user). Fewer than 5 packets in 10^8 should be duplicated.

- *Mis-sequenced*

This parameter is only relevant in relation to packets from the same source to the same destination. Although packets may be interleaved and in transit concurrently, they should arrive in the same sequence in which they are sent; that is, the BOMs should arrive in the same relative order in which they are sent: if not they are mis-sequenced. Fewer than 5 packets in 10^9 should be mis-sequenced.

Delay

The finesses of specifying delay are considerable; but it is basically about how long it takes for a packet to be transferred from one SNI to another. Delay clearly depends on the access rates. For individually addressed packets with up to 9188 octets of user information the following delay objectives are specified:

DS3 SNI ↔ DS3 SNI	95% of packets delivered in less than 20 ms
DS3 SNI ↔ DS1 SNI	95% of packets delivered in less than 80 ms
DS1 SNI ↔ DS1 SNI	95% of packets delivered in less than 140 ms

The same objectives apply to equivalent combinations of E1 and E3 accesses. For group addressed packets the delay objectives are 80 ms higher.

SMDS networks are still somewhat immature and it remains to be seen how well these performance and quality of service objectives are actually met. The prospective customer should investigate the service provider's track record before investing.

4.3 EARLY SMDS IMPLEMENTATIONS

Introducing any new service brings the problem that CPE manufacturers do not want to put money into developing the necessary new terminal equipment until they are sure of the market, and service providers have difficulty selling a service for which little or no CPE is available. The CPE manufacturers say there is no demand, the service providers say that is because there is no CPE!

This problem was foreseen for SMDS and the choice of an already developed protocol, the IEEE's DQDB protocol for MANs, was made partly to get around it. Nevertheless, the DQDB-capable CPE needed to support multiple-CPE configurations is comparatively expensive, and many customers will want to try the service out before making large-scale investments in it. Suppliers naturally look for ways of getting new kit on to the market at the earliest opportunity and with a minimum of development.

Put these factors together and it is not surprising that early implementations of SMDS focus mainly on offering the single-CPE configuration which demands much less functionality in the CPE. Furthermore, small niche suppliers are often more responsive to new opportunities than larger established manufacturers who may take a less speculative view of the market. The result has been the early availability from specialist companies of low-cost adapters that provide the single-CPE DQDB capability external to the CPE proper (hosts, routers, etc), minimising the development needed in the CPE.

The SMDS data exchange interface (DXI)

These adapters, usually known as Data Service Units or DSUs, exploit the fact that virtually all pre-SMDS CPE already has an HDLC frame-based interface. An SMDS Data eXchange Interface has been defined, generally referred to as DXI, that uses an HDLC frame-based level 2 protocol between the CPE and the DSU, as shown in Figure 4.13. The DSU converts this to the DQDB slot-based level 2 defined for SMDS proper. The SIP level 3 (a purely software process) resides in the CPE.

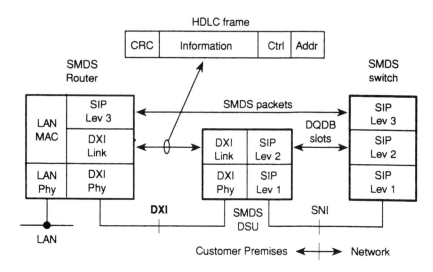

Figure 4.13 SMDS data exchange interface (DXI)

The data exchange interface passes complete SMDS packets to the DSU in the information field of simple frames with the format shown in Figure 4.13 using the HDLC protocol for connectionless data transfer. This is a very simple protocol that can transfer frames from CPE to DSU, and from DSU to CPE. Each frame may carry an SMDS packet, a link management message, or a 'heartbeat' test message.

Though not shown, there is a two-octet header attached to the SMDS packet before inserting it into a frame's information field. This header may be used by the DSU to indicate to the CPE that it is congested or temporarily overloaded and that the CPE should back off temporarily. For this purpose the frame may even contain a null information field.

Link management messages carry information in support of the SMDS DXI Local Management Interface (LMI) procedures, which are not described here.

The 'heartbeat' procedure enables either the CPE or the DSU to periodically check the status of the link connecting them by sending a 'heartbeat' message and checking that an appropriate response is received within a time-out period, typically 5 seconds.

The DXI/SNI

More recently a DXI/SNI interface has been defined, based on the DXI, in which the simple HDLC-based level 2 protocol actually operates directly between the CPE and the SMDS switch, as shown in Figure 4.14, so that the DSU is not needed. This acknowledges the belief that many customers will want only the single-CPE capability, and enables the switch manufacturer to exploit the simplicity it brings.

The protocol operated over the DXI/SNI is basically the same as described

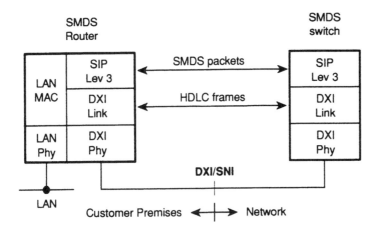

Figure 4.14   SMDS DXI/SNI

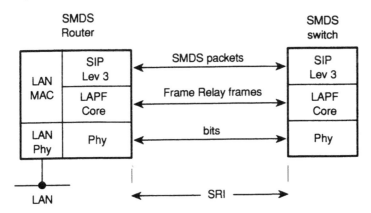

Figure 4.15   SMDS relay interface

above for the DXI, but is designed for low-speed operation at data rates of 56, 64, $N \times 56$ and $N \times 64$ kbit/s.

The SMDS relay interface (SRI)

One of the reasons that new services are costly during the introduction phase is that there are comparatively few point of presence—that is SMDS switches in the case of SMDS—so that the access links between customer and switch tend to be long. Recognising that many CPE manufacturers have already developed Frame Relay interfaces for their products, and that many SMDS customers will be satisfied with sub-2Mbit/s access rates, an alternative SNI has been defined, known as SIP relay interface or SRI, based on Frame Relay as the Level 2 protocol, as shown in Figure 4.15.

SRI uses a Frame Relay PVC between the CPE and the SMDS switch. The data transfer protocol is that described in chapter 3 based on the LAPF core.

Only one PVC is supported on an SRI interface, and no more than one frame may be in transit in each direction of transmission. Like DXI/SNI the SRI is designed for access rates of 56, 64, $N \times 56$ and $N \times 64$ kbit/s.

The DXI/SNI PVC may in principle pass through a Frame Relay network *en route* to the SMDS switch. But there are two important differences between the Frame Relay protocol as defined for DXI/SNI and as defined for the Frame Relay service. Firstly, in order to carry a complete SMDS packet in a single frame, the SRI frame must support an information field length of up to 9232 octets; the maximum for the Frame Relay service is 1600 octets. Secondly, for frames longer than 4096 octets it is recommended that a 32-bit CRC is used, rather than the standard Frame Relay CRC of 16 bits, in order to maintain a satisfactory error detection rate for the longer frames.

Although these differences do not actually prevent us from going through a Frame Relay network to access the SMDS service using the SRI, it remains to be seen whether the 1600-octet constraint on SMDS packet length is acceptable. In view of the success Frame Relay is enjoying for LAN interconnection, despite its 1600 octet limit, it is the authors' view that it is not a problem. Note that the CRC length issue disappears if the SMDS packet length is limited to a maximum of 1600 octets.

4.4 Summary

Aimed primarily at wide area LAN interconnection, SMDS is a connectionless high-speed packet data service with the capability to support virtual private networking for data. It is considered by many to be the first truly broadband wide area service of all, and represents real co-operation between the IT world (which developed the DQDB MAN protocol) and the telecommunications world.

It is designed to be a technology-independent service. Though initial implementation is based on the DQDB MAN protocol, the idea is that the network technology can evolve independently of the service. That this idea works is illustrated by the DXI, DXI/SNI and SRI interfaces, none of which uses MAN technology. There is no reason why CPE using any combination of these network technologies for network access should not interoperate to provide the SMDS service. In due course there is no doubt that SMDS will also be offered over an ATM-based network (see Chapter 5).

So SMDS as a service is comparatively future-proof, and a customer need have little fear that building on SMDS will involve costly upgrading. But it is at the beginning of the take-up curve, and there are likely to be significant differences between the services offered by different network operators, both in terms of features and reliability. So the potential customer should look at the small print and the track record, as well as tariffs, before choosing a supplier.

Using the DXI and SRI interfaces SMDS can be made available at lower access rates, and therefore lower cost. This provides considerable flexibility to meet a wide range of customer requirements, including those who want to try the service out before making large-scale commitment. It creates a clear overlap with Frame Relay. For many customers there will therefore be a choice to make between SMDS and Frame Relay. The descriptions given here and in chapter 3 should provide useful guidance in identifying the important differences and similarities between the two services.

The description has been made in two 'passes'. Section 4.1 gives an introduction, designed for easy understanding, but therefore lacking the detail that some readers want. Section 4.2 adds detail for the more demanding reader. Anyone needing more detail than this is clearly serious about it, and should really get to grips with the source documents identified below.

REFERENCES

General

Byrne, W. R. *et al.* (1991) Evolution of metropolitan area networks to broadband ISDN. *IEEE Communications Magazine*, January.

Fischer, W. *et al.* (1992) From LAN and MAN to broadband ISDN. *Telcom Report International*, **15**, No. 1.

Standards

Bellcore Technical Advisories (to save trees this list is not exhaustive; the Advisories listed form a good starting point for the serious reader)

TR-TSV-000772 Generic Systems requirements in support of Switched Multi-Megabit Data Service

TR-TSV-000773 Local Access System Generic Requirements, Objectives and Interfaces in support of Switched Multi-Megabit Data Service

TR-TSV-000774 SMDS Operations Technology Network Element Generic Requirements

TR-TSV-000775 Usage Measurement Generic Requirements in support of Billing for Switched Multi-Megabit Data Service

ITU-TS

E.164 Numbering Plan for the ISDN Era

I.364 Support of Broadband Connectionless Data Service on B-ISDN

F.812 Broadband Connectionless Bearer Service, Service Description

ISO/IEC 8802-6, Distributed Queue Dual Bus (DQDB) access method and physical layer specifications (also known as ANSI/IEEE Std 802.6)

SMDS Interest Group (SIG) and European SMDS Interest Group (ESIG)

SIG-TS-001/1991 Data Exchange Interface Protocol

SIG-TS-002/1992 DXI Local Management Interface

SIG-TS-005/1993 Frame Based Interface Protocol for SMDS Networks—Data Exchange Interface/Subscriber Network Interface

SIG-TS-006/1993 Frame Based Interface Protocol for Networks Supporting SMDS—SIP Relay Interface

ESIG-TS-001/92 SMDS Subscriber Network Access Facility Service and Level 2 and 3 Subscriber Network Interface Specification

ESIG-TI-001/93 CBDS/SMDS Service Definition Comparison

ETSI

ETS 300 211 to 216 Metropolitan Area Network (MAN)

ETS 300 217 Connectionless Broadband Data Service (CBDS)

5

Asynchronous Transfer Mode (ATM)

All this buttoning and unbuttoning

18th century suicide note

The history of telecommunications is basically a history of technology. Advances in technology have led to new networks, each new network offering a range of new services to the user. The result is that we now have a wide range of networks supporting different services. We have the telex network, the telephone network, the ISDN, packet-switched data networks, circuit-switched data networks, mobile telephone networks, the leased line network, local area networks, metropolitan area networks, and so on. More recently we have seen the introduction of networks to support Frame Relay and SMDS services. The problem is that the increasing pace of developments in applications threatens to make new networks obsolete before they can produce a financial return on the investment.

To avoid this problem it has long been the telecommunications engineer's dream to develop a universal network capable of supporting the complete range of services, including of course those that have not yet been thought of. The key to this is a switching fabric flexible enough to cater for virtually any service requirements. ATM is considered by many to be as close to this as we are likely to get in the foreseeable future. This chapter explains the basic ideas of ATM, how it can carry different services in a unified way, and how it will provide seamless networking over both the local and wide areas, i.e. Total Area Networking. Section 5.1 gives a general overview of the key features of ATM with an explanation of the underlying principles. Section 5.2 puts a bit more flesh on the skeleton. Section 5.3 looks at how SMDS and Frame Relay are carried over an ATM network, and section 5.4 looks briefly at ATM in local area networks.

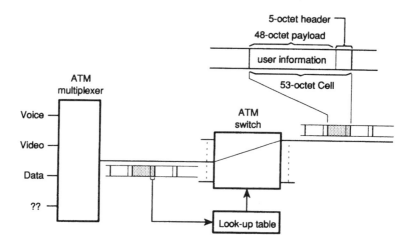

Figure 5.1 ATM cells: the universal currency of exchange

5.1 THE BASICS OF ATM

Cell switching

The variety of networks has arisen because the different services have their own distinct requirements. But despite this variety, services can be categorised broadly as continuous bit-stream oriented, in that the user wants the remote end to receive the same continuous bit-stream that is sent; or as bursty, in that the information generated by the user's application arises in distinct bursts rather than as a continuous bit-stream. Generally speaking, continuous bit-stream oriented services map naturally on to a circuit-switched network, whereas bursty services tend to be better served by packet-switched networks. Any 'universal' switching fabric therefore needs to combine the best features of circuit-switching and packet-switching, while avoiding the worst.

There is also great diversity in the bit rates that different services need. Interactive screen-based data applications might typically need a few kilobits per second. Telephony needs 64 kbits/s. High-quality moving pictures may need tens of megabits per second. Future services (such as holographic 3D television or interactive virtual reality) might need many tens of megabits per second. So the universal network has to be able to accommodate a very wide range of bit rates.

The technique that seems best able to satisfy this diversity of needs is what has come to be called cell-switching which lies at the heart of ATM. In cell-switching the user's information is carried in short fixed-length packets known as cells. As standardised for ATM, each cell contains a 5-octet header and a 48-octet information field, as shown in Figure 5.1. On transmission links, both between the user and the network and between switches within the network, cells are transmitted as continuous streams with no intervening

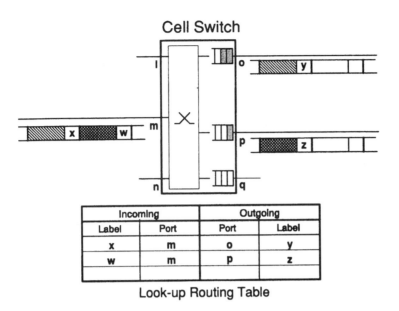

Cell Switch

Incoming		Outgoing	
Label	Port	Port	Label
x	m	o	y
w	m	p	z

Look-up Routing Table

Figure 5.2 Cell switching

spaces. So if there is no information to be carried, empty cells are transmitted to maintain the flow.

User information is carried in the information field, though for reasons that will become clear the payload that a cell carries is sometimes not quite 48-octets. The cell header contains information that the switches use to route the cell through the network to the remote terminal. Because it is only 5-octets long the cell header is too short to contain a full address identifying the remote terminal and it actually contains a label that identifies a connection. So cell-switching, and therefore ATM, is intrinsically connection-oriented (we will see later how connectionless services can be supported by ATM). By using different labels for each connection a terminal can support a large number of simultaneous connections to different remote terminals. Different connections can support different services. Those requiring high bit-rates (such as video) will naturally generate more cells per second than those needing more modest bit rates. In this way ATM can accommodate very wide diversity in bit rate.

The basic idea of ATM is that the user's information, after digital encoding if not already in digital form, is accumulated by the sending terminal until a complete cell payload has been collected, a cell header is then added and the complete cell is passed to the serving local switch for routing through the network to the remote terminal. The network does not know what type of information is being carried in a cell; it could be text, it could be speech, it could be video, it might even be telex! Cell switching provides the universal switching fabric because it treats all traffic the same (more or less—read on for more detail), whatever service is being carried.

Figure 5.2 illustrates the principle of cell switching. A number of transmission links terminate on the cell switch, each carrying a continuous stream of cells.

All cells belonging to a particular connection arrive on the same transmission link and are routed by the switch to the desired outgoing link where they are interleaved for transmission on a first-come-first-served basis with cells belonging to other connections. For simplicity only one direction of transmission is shown. The other direction of transmission is treated in the same way, though logically the two directions of transmission for a connection are quite separate. Indeed, as we shall see, one of the features of ATM is that the nature of the two channels forming a connection (one for each direction of transmission) can be configured independently.

Following usual packet-switching parlance, ATM connections are more correctly known as 'virtual' connections to indicate that, in contrast with real connections, a continuous end-to-end connection is not provided between the users. But to make for easier reading in what follows, the term 'virtual' is generally omitted; for connection read virtual connection.

A connection is created through the network by making appropriate entries in routing look-up tables at every switch *en route*. This would be at subscription time for a permanent virtual circuit (PVC) or call set-up time for a switched virtual circuit (SVC) (for simplicity here, aspects of signalling are omitted). Each (horizontal) entry in the routing look-up table relates to a specific connection and associates an incoming link and the label used on that link to identify the connection with the desired outgoing link and the label used on that link to identify the connection. Note that different labels are used on incoming and outgoing transmission links to identify the same connection (if they happen to be the same it is pure coincidence).

Figure 5.2 shows successive cells arriving on incoming link **m**, each associated with a different connection, i.e. they have different labels on that link. The routing table shows that the cell with incoming label **x** should be routed out on link **o**. It also shows that the label to be used for this connection on outgoing link **o** is **y**. Similarly, the incoming cell with label **w** should be routed out on link **p**, with the new label **z**. It is clear therefore that different connections may use the same labels, but not if they are carried on the same transmission link.

Because of the statistical nature of traffic, no matter how carefully designed an ATM network is, there will be occasions (hopefully rare) when resources (usually buffers) become locally overloaded and congestion arises. In this situation there is really no choice but to throw cells away. To increase the flexibility of ATM, bearing in mind that some services are more tolerant of loss than others, a priority scheme has been added so that when congestion arises the network can discard cells more intelligently than would otherwise be the case. There is a single bit in the cell header (see Figure 5.10) known as the Cell Loss Priority bit (CLP) that gives an indication of priority. Cells with CLP set to 1 are discarded by the network before cells with CLP set to 0. As will be seen, different cells belonging to the same connection may have different priority.

Choice of cell length

If cells are going in the same direction, the switch may route them simultaneously to the same outgoing link. Since only one cell can actually be transmitted at a time, it is necessary to include buffer storage to hold contending cells until they can be transmitted. Contending cells queue for transmission on the outgoing links. By choosing a short cell length, the queueing delay that is incurred by cells *en route* through the network can be kept acceptably short.

Another important consideration that favours a short cell length is the time it takes for a terminal to accumulate enough information to fill a cell, usually referred to as the packetisation delay. For example, for a digital telephone which generates digitally encoded speech at 64 kbit/s it takes about 6 ms to fill a cell. This is delay that is introduced between the speaker and the listener, additional to any queueing delays imposed by the network. Speech is particularly sensitive to delay because of the unpleasant effects of echo that arise when end-to-end delays exceed about 20 ms.

One of the important effects caused by queueing in the network is the introduction of cell delay variation in that not all cells associated with a particular connection will suffer the same delay in passing through the network. Although cells may be generated by a terminal at regular intervals (as for example for 64 kbit/s speech) they will not arrive at the remote terminal with the same regularity. Some will be delayed more than others. To reconstitute the 64 kbit/s speech at the remote terminal a reconstruction buffer is needed to even out the variation in cell delay introduced by the network. This buffer introduces yet more delay, often referred to as depacketisation delay. Clearly, the shorter the cell the less cell delay variation there will be and the shorter the depacketisation delay.

So the shorter the cell the better. But this has to be balanced against the higher overhead which the header represents for a shorter cell length, and the 53-octet cell has been standardised for ATM as a compromise. The saga of this choice is interesting and reflects something of the nature of international standardisation. Basically, Europe wanted very short cells with an information field of 16 to 32 octets so that speech could be carried without the need to install echo suppressors, which are expensive. The USA on the other hand wanted longer cells with a 64 to 128 octet information field to increase the transmission efficiency; the transmission delays on long distance telephone circuits in the USA meant that echo suppressors were commonly fitted anyway. CCITT (now ITU-TS) went halfway and agreed an information field of 48 octets, thought by many to combine the worst of both worlds!

Network impairments

The dynamic allocation of network resources inherent in cell-switching brings the flexibility and transmission efficiencies of packet switching, whereas the short delays achieved by having short fixed-length cells tend towards the more predictable performance of circuit switching. Nevertheless, impairments do arise in the network, as we have seen, and they play a central role in service definition and network design, as we shall see. The main impairments are as follows.

- Delay: especially packetisation delay, queueing delay, and depacketisation delay, though additionally there will be switching delays and propagation delay.

- Cell delay variation: different cells belonging to a particular virtual connection will generally suffer different delay in passing through the network because of queueing.

- Cell loss: may be caused by transmission errors that corrupt cell headers, congestion due to traffic peaks or equipment failure

- Cell misinsertion: corruption of the cell header may cause a cell to be routed to the wrong recipient. Such cells would be lost to the intended recipient and inserted into the wrong connection.

Control of these impairments in order to provide an appropriate quality of service over a potentially very wide range of services is one of the dominating themes of ATM.

The traffic contract

From what we have seen of cell-switching so far it should be clear that new connections would compete for the same network resources (transmission capacity, switch capacity and buffer storage) as existing connections. It is important, therefore, to make sure that creating a new connection would not reduce the quality of existing connections below what is acceptable to the users. But what is acceptable to users? We have seen that one of the key attractions of ATM is its ability to support a very wide range of services. These will generally have different requirements, and what would be an acceptable network performance for one service may be totally unacceptable for another. Voice, for example, tends to be more tolerant to cell loss than data, but much less tolerant to delay. Furthermore, for the network to gauge whether it has the resources to handle a new connection it needs to know what the demands of that connection would be.

A key feature of ATM is that for each connection a traffic contract is agreed between the user and the network. This contract specifies the characteristics of the traffic the user will send into the network on that connection and it specifies the quality of service (QoS) that the network must maintain. The contract places an obligation on the network; the user knows exactly what service he is paying for and will doubtless complain to the service provider if he does not get it. And it places an obligation on the user; if he exceeds the agreed traffic profile the network can legitimately refuse to accept the excess traffic on the grounds that doing so may compromise the quality of the existing connections and thereby breach the contracts already agreed for those connections. But provided that the user stays within the agreed traffic profile the network should support the quality of service requested. The contract also provides the basis on which the network decides whether it has the resources available to support a new connection; if it does not then the new connection request is refused.

Traffic characteristics are described in terms of parameters such as peak cell rate, together with an indication of the profile of the rate at which cells will be sent into the network. The quality of service is specified in terms of parameters relating to accuracy (such as cell error ratio), dependability (such as cell loss ratio), and speed (such as cell transfer delay and cell delay variation). Some of these parameters are self-explanatory, some are not. They are covered in more detail later, but serve here to give a flavour of what is involved.

We may summarise this as follows:

- for each connection the user indicates his service requirements to the network by means of the traffic contract;

- at connection set-up time the network uses the traffic contract to decide, before agreeing to accept the new connection, whether it has the resources available to support it while maintaining the contracted quality of service of existing connections; the jargon for this is connection admission control (CAC);

- during the connection the network uses the traffic contract to check that the users stay within their contracted service; the jargon for this is usage parameter control (UPC).

How this is achieved is considered in more detail in section 5.2.

Adaptation

ATM, then, offers a universal basis for a multiservice network by reducing all services to sequences of cells and treating all cells the same (more or less). But, first we have to convert the user's information into a stream of cells, and of course back again at the other end of the connection. This process is known as ATM adaptation, and is easier said than done! The basic idea behind adaptation is that the user should not be aware of the underlying ATM network infrastructure (we will look at exceptions to this later when we introduce native-mode ATM services).

Circuit emulation—an example of ATM adaptation

For example, suppose that the user wants a leased-line service; this should appear as a direct circuit connecting him to the remote end, i.e. the ATM

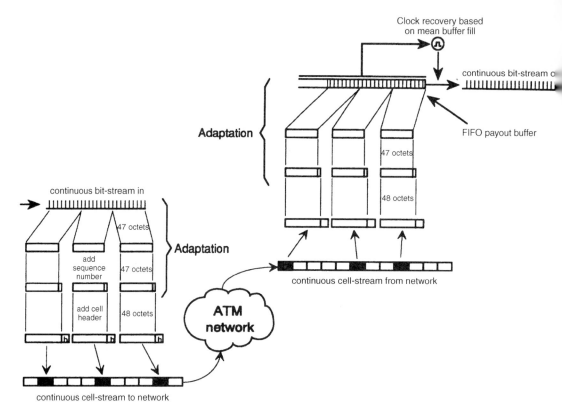

Figure 5.3 ATM adaptation for circuit emulation

network should emulate a real circuit. The user transmits a continuous clocked bit stream, at say 256 kbit/s, and expects that bit stream to be delivered at the remote end with very little delay and with very few transmission errors (and similarly in the other direction of transmission). As shown in Figure 5.3, at the sending end the adaptation function would divide the user's bit-stream into a sequence of octets. When 47 octets of information have been accumulated they are loaded into the information field of a cell together with a one-octet sequence number. The appropriate header is added, identifying the connection, and the cell is sent into the network for routing to the remote user as described above. This process of chopping the user information up so that it fits into ATM cells is known as segmentation.

At the receiving end the adaptation function performs the inverse operation of extracting the 47 octets of user information from the cell and clocking them out to the recipient as a continuous bit-stream, a process known as re-assembly. This is not trivial. The network will inevitably have introduced cell delay variation, which will have to be compensated for by the adaptation process. The clock will have to be recreated so that the bit stream can be clocked out to the recipient at the same rate at which it was input by the sender. The one-octet sequence number sent in every cell allows the terminating equipment to detect whether any cells have been lost in transit through the network (not that anything can be done in this case to recover the lost information but the loss can be signalled to the application).

To overcome the cell delay variation the adaptation process will use a re-assembly buffer (sometimes called the play-out buffer). The idea is that the re-assembly buffer stores the payloads of all the cells received for that connection, for a period equal to the maximum time a cell is expected to take to transit the network, which includes cell delay variation. This means that if the information is clocked out of the re-assembly buffer at the same clock rate as the original bit stream (256 kbit/s in this example) the re-assembly buffer should never empty and the original bit stream would be recreated (neglecting any loss of cells).

The re-assembly buffer would also be used to recreate the play-out clock. Typically a phase-locked loop would be used to generate the clock. The fill level of the buffer, i.e. the number of cells stored, would be continously compared with the long-term mean fill level to produce an error signal for the phase-locked loop to maintain the correct clock signal.

It is clear from this simple (and simplified!) example that the adaptation process must reflect the nature of the service to be carried and that a single adaptation process such as that outlined above will not work for all services. But it is equally clear that having a different adaptation process for every possible application is not practicable. CCITT has defined a small number of adaptation processes, four in all, each applicable to a broad class of services having features in common. The example shown above (circuit emulation) could be used to support any continuous bit rate service, though the bit rate and quality of service needed would depend on the exact service required.

The ATM protocol reference model

A layered reference model for ATM has been defined as a framework for the detailed definition of standard protocols and procedures, as shown in Figure 5.4 (I.321). There are essentially three layers relating to ATM: the physical layer; the ATM layer; and the ATM adaptation layer. (It should be noted that these layers do not generally correspond exactly with those of the OSI 7-layer model.) Each of the layers is composed of distinct sublayers, as shown.

Management protocols, not shown, are also included in the reference model, for both layer management and plane management. For the sake of brevity these are not covered here.

The physical layer

The physical layer is concerned with transporting cells from one interface through a transmission channel to a remote interface. The standards embrace a number of types of transmission channel, both optical and electrical, including SDH (synchronous digital hierarchy) and PDH (plesiochronous

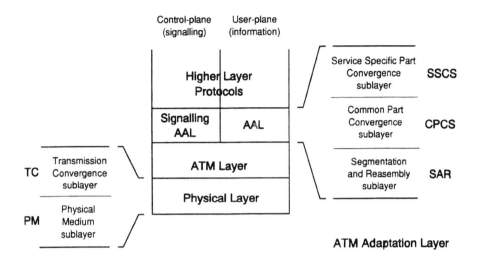

Figure 5.4 The ATM protocol reference model

digital hierarchy). The physical layer may itself generate and insert cells into the transmission channel, either to fill the channel when there are no ATM cells to send or to convey physical layer operations and maintenance information: these cells are not passed to the ATM Layer. The physical layer is divided into a physical medium (PM) sublayer, which is concerned only with medium-dependent functions such as line coding, and a transmission convergence (TC) sublayer, which is concerned with all the other aspects mentioned above of converting cells from the ATM layer into bits for transmission, and vice versa for the other direction of transmission.

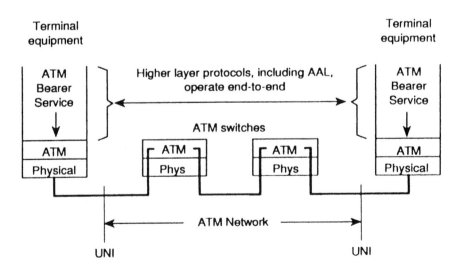

Figure 5.5 ATM bearer service

The ATM Layer (I.361)

The ATM layer is the layer at which multiplexing and switching of cells take place. It provides virtual connections between end-points and maintains the contracted quality of service by applying a connection admission control procedure at connection set-up time and by policing the agreed traffic contract while the connection is in progress. The ATM layer provides to higher layers a service known as the ATM bearer service, as shown in Figure 5.5.

The ATM adaptation layer (AAL) (I.363)

The ATM adaptation layer, invariably referred to simply as the AAL, translates between the service required by the user (such as voice, video, Frame Relay, SMDS, X.25) and the ATM bearer service provided by the ATM layer. It is composed of the convergence sublayer (CS) and the segmentation and reassembly sublayer (SAR). The convergence sublayer performs a variety of functions which depend on the actual service being supported, including clock recovery, compensating for cell delay variation introduced by the network, and dealing with other impairments introduced by the network such as cell loss. The segmentation and reassembly sublayer segments the user's information, together with any supporting information added by the convergence sublayer, into blocks that fit into the payload of successive ATM cells for transport through the network, and in the other direction of transmission it reassembles cells received from the network to recreate the user's information as it was before segmentation at the sending end.

Four types of AAL have been defined. To further minimise the variety the AAL convergence sublayer is itself divided into a common part (CPCS) and a Service Specific part (SSCS). For each type of AAL the common part deals with those features that the supported services have in common, whereas the service specific part deals with things that are different. The AALs are considered in more detail in section 5.2.

As Figure 5.4 shows, different AALs are used for signalling and for the data path; that is the control plane and user plane in CCITT parlance. In effect signalling is viewed as a special type of service and a signalling AAL (SAAL) has been developed to support it.

The ATM adaptation layer is not mandatory for the data path (i.e. the user plane), and may be omitted. Applications that can use the ATM bearer service as provided by the ATM Layer may do so. Indeed, it seems likely that in the fullness of time, when ATM is common and ubiquitous, applications will be designed to use the ATM bearer service directly rather than via an intermediate service such as Frame Relay. In the case of permanent virtual connections there is no requirement for user signalling, so the signalling AAL may also be missing.

Virtual paths and virtual circuits

We now take the story of cell-switching a bit further. So far we have considered that a straightforward virtual connection is created between the

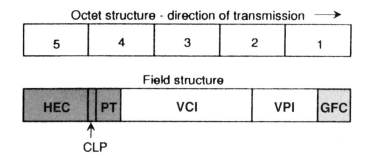

Figure 5.6   The ATM cell header

Logical structure

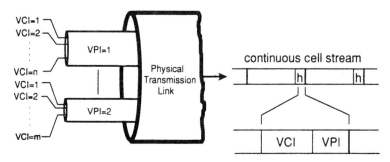

Figure 5.7   Virtual paths and virtual channels

end users, and that labels are used in the cell headers so that each cell can be associated with the appropriate connection. In fact for ATM two types of virtual connection have been defined the virtual path connection (VPC) and the virtual channel connection (VCC), and the label actually consists of two distinct parts, as shown in Figure 5.6: a virtual path identifier (VPI) and the virtual channel identifier (VCI) (we will look at the other header fields in section 5.2).

A virtual path connection is a semi-permanent connection which carries a group of virtual channels all with the same end-points. On any physical link the VPIs identify the virtual paths, and the VCIs identify the virtual channels, as shown in Figure 5.7. VPIs used on one physical link may be re-used on others, but VPIs on the same physical link are all different. A VCI relates to a virtual path; so different virtual paths may re-use VCIs, but VCIs on the same virtual path are all different.

Virtual paths and virtual circuits both have traffic contracts. These are notionally independent, and different virtual channels in the same virtual path may have different qualities of service. But the quality of service of a virtual channel cannot be better than that of the virtual path in which it is carried.

Perhaps the easiest way to explain the idea of Virtual Paths is to look at a few examples of how they might be used.

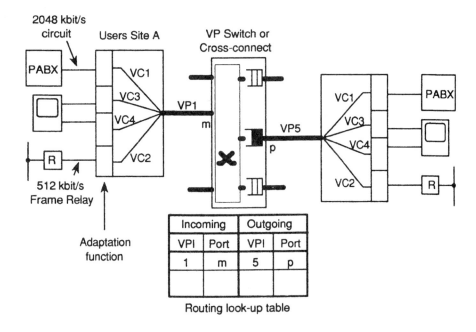

Figure 5.8 A virtual path connection between user sites

Virtual path example 1—flexible interconnection of user sites

Companies often want to interconnect geographically remote sites. If conventional leased lines are used for this, several of them are usually needed, typically involving a mix of bit rates and qualities. Some may be used to interconnect the company's PABXs. Others might be used to interconnect LAN routers, or to support videoconferencing, or whatever.

With ATM the company could lease from the network operator a single virtual path connection between the two sites, as shown in Figure 5.8.

Each site has a PABX for telephony, a local area network, and a videoconferencing facility. The virtual path supports four virtual channels with VCI = 1, 2, 3 and 4. At each site an ATM multiplexer performs the ATM adaptation function.

The virtual channel with VCI = 1 carries a circuit-emulation service, in effect interconnecting the PABXs by a 2048 kbit/s private circuit.

The virtual channel with VCI = 2 carries a Frame Relay service at 512 kbit/s interconnecting the two LANs via the routers.

The virtual channel with VCI = 3 carries 64 kbit/s constant bit rate voice for the videoconferencing facility, and the fourth, with VCI = 4, carries constant bit rate video at 256 kbit/s also for the videoconferencing facility.

The virtual path connection is set up on a subscription basis by the network operator via virtual path cross-connection switches in the ATM network, only one of which is shown for simplicity. The virtual path cross-connection switches are simple cell-switches and operate as decribed in Figure 5.2. But, in this case the switches look only at the VPIs: they do not look at the VCIs which are transported through the virtual path connection unchanged and unseen by the network. The routing look-up table therefore has entries relating only to VPIs. In this case all cells incoming on port m with VPI = 1 are routed out on port p with new VPI = 5. The VCIs are unchanged by the VP cross-connect and are the same both ends.

This use of a virtual path simplifies the network switching requirements substantially, since the virtual channels do not have to be individually switched. It gives the user great flexibility to use the capacity of the virtual path in any way he wishes, since the virtual channel structure within the virtual path is not seen by the network. The user could, for example, set up other virtual channels between the ATM multiplexers to interconnect other devices such as surveillance cameras for remotely checking site security. Or virtual channels could be allocated different bit rates at different times of the day to exploit the daily variations in traffic. But the user must make sure that the aggregate demand of the virtual channels does not exceed the traffic contract agreed for the virtual path.

Depending on tariffs and service requirements, it may of course be better for the company to lease several virtual paths between the two sites, each configured to carry virtual channels needing a particular quality of service. The requirements of voice, as reflected in the circuit emulation service connecting the two PABXs, are quite different from LAN interconnection requirements. So it might be beneficial to use different virtual paths for these.

Virtual path example 2—flexible network access

The above example is a bit unrealistic in that it only provides for traffic between the two sites. In practice a great deal of the traffic, especially voice traffic, would need to be routed to third parties. Figure 5.9 shows how an additional virtual path (VPI = 2) is used to provide PABX access to the PSTN.

Figures 5.8 and 5.9 show only one direction of transmission. The other direction of transmission is treated in an identical way. It should be noted, however, that although connections generally involve two-way transmission, that is a channel in each direction, this is not mandatory: one-way connections are permitted.

In a practical network virtual path cross-connection switching and virtual channel switching are likely to be combined in one switching system. This would give even greater flexibility than shown above. For example, the user's traffic could be segregated by the ATM multiplexer into voice and data, all data being carried on one virtual path, and all voice traffic being carried on a second, whether intended for the remote site or the PSTN. The combined VP/VC switch could then route the voice traffic as required.

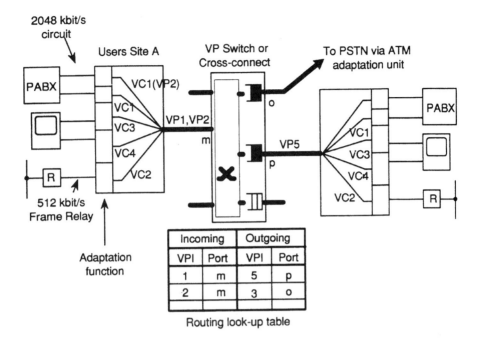

Figure 5.9 Flexible access using virtual paths

Virtual path example 3—Using virtual paths within the ATM network

The distinction between virtual paths and virtual channels can substantially simplify switching and multiplexing in the network, and adds an important degree of freedom to network designers and operators. In effect virtual paths can be used to create a logical network topology that is quite distinct from that of the physical links. A virtual path can interconnect two switches even if there is no direct link between them. For routing purposes the virtual path constitutes a direct connection. Virtual paths can also be used between switches as a way of logically partitioning different types of traffic that need different qualities of service. This can significantly simplify connection admission control schemes.

5.2 COMPLETING THE PICTURE

The above account is intended to provide an accessible picture of ATM (basically what it is, and perhaps what it is not) and the emphasis has been on clarity of explanation rather than detail. For some readers this is enough; they can pass on to other chapters. Other readers will want more, and they should plough on.

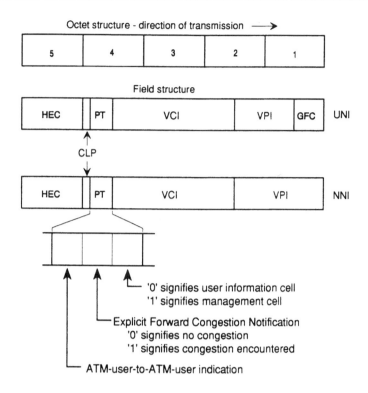

Figure 5.10 The ATM cell header

The ATM cell

We begin by revisiting the cell header. So far we have looked at the function of the VPI, the VCI and the CLP bit. Now we will look at the others.

There are in fact two slightly different cell header formats, one used at the User Network Interface (UNI), the other at the Network–Network Interface (NNI) as shown in Figure 5.10. They differ only in that the UNI format includes an additional field, the generic flow control field (GFC), which the NNI format does not. The NNI format takes advantage of the available space to have a longer VPI.

The generic flow control field is intended to provide a multiple access capability (similar to the MAC Layer in LANs) whereby a number of ATM terminals and devices can be attached to a single network interface, each getting access to set-up and clear connections and transfer data on these connections in a standardised and controlled way. At the time of writing the details of this multiple access scheme have not been formally agreed, but are likely to be based on the Orwell protocol developed at BT Laboratories. Clearly there is no requirement for this feature at the NNI, and the NNI format fills the first four bits of the cell header with an extention of the VPI field, permitting more virtual paths to be supported.

The 3-bit PT field indicates the type of payload being carried by the cell. There are basically two types of payload: those which carry user information

Feature	Service Class			
	A	B	C	D
Timing	Required		Not required	
Bit-rate	Constant	Variable		
Connection mode	Connection-oriented			CLS

Figure 5.11 *Service classes*

and those which do not. Cells carrying user information are identified by having 0 in the most significant bit of the PT field. (Strictly speaking there are six cases where this is not true: they are identified by virtue of having specific VCI values that are reserved for signalling or management purposes and not available to carry user information.) Cells with 1 in the most significant bit of the PT field carry OAM or resource management information. We do not consider them further here (the interested reader should consult the standards (I.610 and I.371)).

In user information cells the middle bit of the PT field is a congestion indication bit: 0 signifies that congestion has not been encountered by the cell; 1 indicates that the cell has actually experienced congestion. The least significant bit of the PT field in a user information cell is the ATM-user-to-ATM-user indication, which is passed unchanged by the network and delivered to the ATM Layer user; that is, the AAL, at the other end of the connection. In the next section we will see an example of the ATM-user-to-ATM-user indication being used by one of the AALs.

The last octet of the cell header contains the header error control (HEC). This is used to check whether the cell header has been corrupted during transmission (it is also used by the physical layer to detect cell boundaries). The error checking code used is capable of correcting single bit errors and detecting multiple bit errors. Cells that are received with uncorrectable errors in the header are discarded.

Services and adaptation (I.363)

To keep the number of adaptation algorithms to a minimum four distinct classes of service have been defined, designated A, B, C and D (I.362). As shown in Figure 5.11, they differ in terms of whether they require a strict time relationship between the two ends, whether they are constant bit rate (CBR) or variable bit rate (VBR), and whether they are connection-oriented or connectionless (CLS).

Class A services are constant-bit-rate and connection-oriented, and involve a strict timing relationship between the two ends of the connection. Circuit emulation, as outlined in the previous section, is a good example of a Class A service. PCM-encoded speech is another.

AAL	Service Class
Type 1	A
Type 2	B
Type 3/4	C and D
Type 5	C and D

Figure 5.12 AALs and service classes

Class B services are also connection-oriented with a strict time relationship between the two ends, but have a variable bit rate. The development of variable bit rate services is still in its infancy, but typically they involve coding of voice or video using compression algorithms that try to maintain the quality associated with the peak bit rate of the information while exploiting the fact that for much of the time the actual information rate is a lot less than the peak rate.

Class C services too are connection-oriented, and have a variable bit rate. But they do not involve a strict time relationship between the two ends, and are generally more tolerant of delay than the real-time services of classes A and B. Information is transferred in variable-length blocks (that is, packets or frames from higher-layer protocols). Connection-oriented data services such as Frame Relay and X.25 are examples of class C services. Signalling is another.

Class D services are variable-bit-rate, do not involve a strict time relationship between the two ends, and are connectionless. Again, information is transferred in variable-length blocks. SMDS is perhaps the best-known example.

There are actually eight possible combinations of timing relationship, connection-mode and bit-rate. The other four combinations not covered by classes A to D do not produce viable services. For example, the idea of a connectionless-mode service with a strict timing relationship between the two ends does not really mean anything.

Clearly, services that differ enough to belong to different classes as defined above are likely to need significantly different things from the AAL, and several types of AAL have been defined to support service classes A to D, as shown in Figure 5.12.

Note that it is not quite as straightforward as having a different type of AAL for each of the four service classes, though this was the original intention. What happened was that AAL types 1, 2, 3 and 4 were defined to support service classes A, B, C and D, respectively. But as the AALs were developed it became clear that there was a lot of commonality between AAL3 and AAL4, and it was eventually decided to combine them into a single AAL, now referred to as 3/4. At the same time it was realised that a lot of the services in classes C and D did not need the complexity and associated overheads of AAL3/4 and a simple and efficient adaptation layer was developed to support them. Originally known as SEAL, it has now been standardised as AAL5.

The main provisions of these AALs are outlined below.

AAL1

There are three variations on AAL1 supporting three specific services: circuit transport (sometimes called circuit emulation); video signal transport; and voice-band signal transport. The segmentation and reassembly sublayer functions are the same for all three, but there are differences in what the convergence sublayer does. AAL1 is in principle also applicable to high-quality audio signal transport, but specific provisions for this have not yet been standardised.

Circuit transport can carry both asynchronous and synchronous signals. Asynchronous here means that the clock of the constant bit rate source is not frequency-locked to a network clock, synchronous means that it is. G.702 signals at bit rates up to 34 368 kbit/s are examples of asynchronous signals; I.231 signals at bit rates up to 1920 kbit/s are examples of syncronous signals.

Video signal transport supports the transmission of constant bit rate video signals for both interactive services, which are comparatively tolerant to errors but intolerant to delay, and distribution services, which are less tolerant to errors but more tolerant to delay. As we will see the convergence sublayer (CS) functions needed for the interactive and distributive services are not quite the same.

Voice-band signal transport supports the transmission of A-law and μ-law encoded speech at 64 kbit/s.

Operation of AAL1

For each cell in the send direction the convergence sublayer (CS) accumulates 47 octets of user information which is passed to the SAR sublayer together with a 4-bit sequence number consisting of a 3-bit sequence count and a 1-bit CS indication (CSI). If the constant bit rate signal has a framing structure (such as the 8 kHz structure on an ISDN circuit-mode bearer service) the CS sublayer indicates the frame boundaries to the remote peer CS by inserting a 1-octet pointer as the first octet of the payload of selected segments, and uses the CSI-bit to indicate to the far end that this pointer is present. This reduces the user information payload of these segment to 46 octets. As outlined below, the CSI bit can also be used to carry clock recovery information.

The SAR protects against corruption of the sequence number in transmission by calculating and adding a 4-bit error code designated the sequence number protection (SNP). This error code enables single-bit errors to be corrected and multiple-bit errors to be detected at the far end of the connection. The 47 octets of user information (or the 46 octets of user information plus the 1-octet pointer) together with the octet containing the sequence number and sequence number protection are passed to the ATM Layer as the complete payload for a cell. (Note that for simplicity of illustration Figures 5.13 and 5.14 do not distinguish the CS from the SAR sublayers.)

At the receiving end, shown in Figure 5.14, the AAL receives the 48-octet payload of cells from the ATM layer. The SAR sublayer checks the sequence

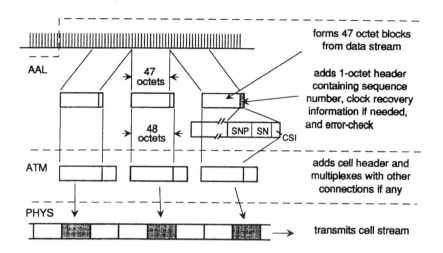

Figure 5.13   AAL1: send direction

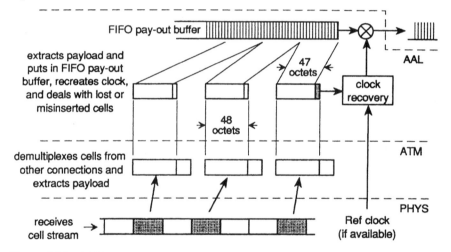

Figure 5.14   AAL1: receive direction

number protection field for errors in the sequence number before passing the sequence number and the associated segment of user information to the CS sublayer. This check would first correct a single-bit error in the sequence number, but if more than one error is detected in the sequence number field the SAR sublayer discards the complete segment and informs the CS sublayer.

The CS sublayer uses the Sequence Numbers to detect the loss or misinsertion of cells. Exactly how it treats these depends on the specific service being carried.

The CS sublayer puts the payload of each segment into the pay-out buffer for clocking out to the user, after extracting any framing pointers if present, the pointers being used to recover and maintain the frame synchronisation of the constant bit rate stream passed to the AAL user.

Two different methods have been identified for clock recovery: the synchronous residual time stamp (SRTS) method and the adaptive clock method.

The SRTS method would be used when a reference clock is available from the network at both ends of the connection. In this method the source AAL CS would periodically send to the remote CS an indication of the difference between the source clock rate and the reference network clock, known as the residual time stamp. The receiving CS uses this information, together with the reference network clock, to keep the pay-out clock correct to within very fine limits. The residual time stamp is carried in the CSI-bit of successive cells (strictly speaking it is carried in the CSI-bit of cells with odd sequence numbers; other information, such as an indication that a structure pointer is present in the payload, would use even-numbered cells).

In the absence of a reference network clock the adaptive clock method would be used. This was outlined in section 5.1 and it uses a locally derived clock whose frequency is controlled by the pay-out buffer fill level. Since the mean data rate of the source bit stream would be reproduced at the receiving end by inserting dummy data into the pay-out buffer to replace any cells lost in transit, the original clock could be recreated within close limits by small variations in the local clock frequency designed to maintain a constant mean buffer fill.

The specific action taken by the AAL CS if cell loss is detected depends on the service being carried. For the circuit-emulation service any dummy bits inserted in the received bit stream in place of lost bits would be set to 1. For the video signal transport service an indication of cell loss could be passed to the AAL user so that appropriate error concealment action could be taken by the video codec, such as repeating the previously received picture frame.

For unidirectional video services (that is video distribution) in which delay is not critical the CS can provide forward error correction. The algorithm prescribed for this can correct up to 4×47 octets (that is 4 segment payloads) in a sequence of 128 cells. But this adds 128 cells-worth of delay and uses the CSI bit which therefore cannot also be used to indicate the bit-stream structure.

AAL2

AAL2, for variable bit rate services, is still being developed, and is too incomplete for inclusion at the time of writing.

AAL3/4

The AAL service specific convergence sublayer (SSCS) deals with adaptation issues that call for different treatment for each of the specific services supported by the AAL. The common part convergence sublayer (CPCS) supports information transfer procedures designed for the broad class of services at which the AAL is aimed (see Figure 5.4). The purpose of this split of functions is simply to minimise the number of AAL variants needed; it makes life easier for implementers, users and standards makers. To make life easier for the reader the following descriptions do not include the service specific convergence sublayer. The important ideas of ATM adaptation are

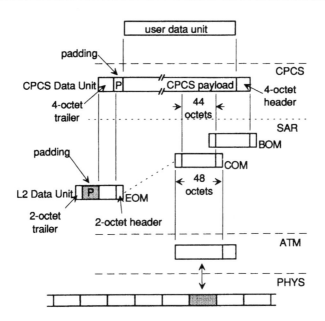

Figure 5.15 AAL3/4

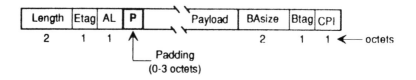

Figure 5.16 AAL3/4 CPCS data unit

covered, but we avoid getting bogged down in spurious detail. We suggest the interested reader investigates service-specific aspects on a need-to-know basis.

AAL3/4 CPCS supports the transfer of variable-length data blocks of up to 65 535 octets between users; for the sake of clarity, in what follows these will be called user data units. Typically these would be packets or frames produced by higher-layer protocols. One of the key features of this sublayer is that it can support the simultaneous transfer of a large number of user data units over a single ATM connection by interleaving their cells.

Figure 5.15 shows how user information is transported by AAL3/4 together with the formats of the data units involved.

The CPCS pads the user data unit so that it is a multiple of 32 bits (for ease of processing) and adds a 4-octet header and 4-octet trailer to form what is here called a CPCS data unit. As Figure 5.16 shows, the CPCS header and trailer include identifiers (Btag and Etag) and length indicators (BAsize and Length) that help the receiving end to allocate appropriate buffer space, reassemble the CS data unit, and do basic error checks.

The same value is inserted in both Btag and Etag fields of a CPCS data unit,

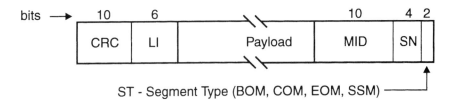

bits →

10	6		10	4	2
CRC	LI	Payload	MID	SN	

ST - Segment Type (BOM, COM, EOM, SSM)

Figure 5.17 AAL3/4 SAR data unit

enabling the remote CPCS to check that the right CPCS header is associated with the right trailer. Different Btag/Etag values are used in each successive CPCS data unit sent. BAsize indicates to the receiving CPCS how much buffer space should be reserved for the associated data unit: specifically it indicates the length of the CPCS data unit payload. CPI indicates the units associated with BAsize (currently restricted to octets). AL is included simply to give 32-bit alignment to the CPCS trailer, and it contains no information.

This CPCS data unit is segmented into 44-octet segments by the SAR sublayer, which adds a 2-octet header and 2-octet trailer to each of the segments, forming what is referred to here as the SAR data unit, illustrated in Figure 5.17. The SAR header indicates whether the SAR data unit is the first segment of the CS data unit, shown as BOM (beginning of message); or an intermediate segment, shown as COM (continuation of message); or the last segment, shown as EOM (end of message). The SAR header also includes a 4-bit sequence number so that the far end can check whether any SAR data units have been lost during transmission.

If the user data unit is short enough to fit into a single SAR data unit (it must be no longer than 36 octets for this since the CPCS adds 8 octets and the SAR sublayer adds 4 octets, and it must fit into the 48 octet payload of a cell) then the SAR data unit is marked as a single segment message (SSM) and is sent as a single cell message.

Since the CPCS data unit is generally not a multiple of 44 octets, end of message and single segment message SAR data units may be only partially filled. The SAR data unit trailer therefore includes a 6-bit length indication (LI) identifying the number of octets of information contained in the payload (the rest is padding).

The SAR data unit trailer also includes a 10-bit cyclic redundancy check that is used to detect whether the SAR data unit header, payload or length indication has been corrupted during transmission. The 48-octet SAR data unit is passed to the ATM layer which adds the appropriate cell header for transmission by the physical layer.

At the receiving end the reverse process takes place. The ATM layer passes the cell payload to the SAR sublayer which immediately checks for transmission errors by comparing a locally generated error check sum with that carried in the SAR data unit trailer. If it is corrupted the data unit is discarded, otherwise its sequence number is checked to make sure that it is the one expected; that is, none have been lost by the network. In the event of an error the SAR sublayer tells the CPCS and the transfer of the CPCS data unit is aborted.

In the absence of errors the SAR sublayer strips off the header and trailer and passes the SAR data unit's payload to the CPCS which adds it to the partially assembled CPCS data unit. If it is the last SAR data unit in the sequence, indicated by an EOM segment type code in the header, the padding will also be stripped off before passing the payload to the CPCS.

On receipt of the last segment the CPCS checks that the CPCS data unit header and trailer correspond (that is, they contain the same reference number (Btag = Etag)) and that the data unit is the right length. It then strips off any padding and passes the payload to the CPCS user as the user data unit. If any of these checks fail the data unit is either discarded or is passed to the user with the warning that the data may be corrupted.

As described above, this process transfers a variable length user data unit over the ATM network and, in the absence of cell loss or corruption, delivers it unchanged to the user at the other end. But there is a little bit more to add since we have not yet explained how a number of user data units may be transferred concurrently.

This is achieved very simply by including a 10-bit multiplexing identifier, MID, in every SAR data unit, as shown in Figure 5.17. A new MID value is allocated to each CPCS data unit as it is created for transmission; the MID value inserted into each SAR data unit then tells the remote CPCS which CPCS data unit the SAR data unit belongs to. Typically the MID value would be incremented for each CPCS data unit sent.

AAL5

AAL5 provides a similar data transport service to AAL3/4, though it does not include a multiplexing capability and can only transfer one CS data unit at a time. However, it provides the service in a much simpler way and with significantly fewer overheads.

Error detection in AAL5 is done entirely in the CPCS so that the SAR sublayer can be very simple. As Figure 5.18 shows, CPCS takes the user data unit, adds an 8-octet trailer (there is no CPCS header) and pads the resulting CPCS data unit out so that it is a multiple of 48 octets. Padding the CPCS data unit in this way avoids the need to pad SAR data units.

As shown in Figure 5.19 the CPCS trailer consists of a 1-octet CPCS user-to-user information field, a 1-octet common part indicator (CPI) field, a 2-octet length field, and a 4-octet cyclic redundancy code (CRC) field for error checking.

The CPCS user-to-user information field carries up to 8 bits of information received from the service specific convergence sublayer (or the CPCS user if there is no SSCS) and transports it transparently to the remote SSCS (or CPCS user).

The common part indicator is really included to give a degree of

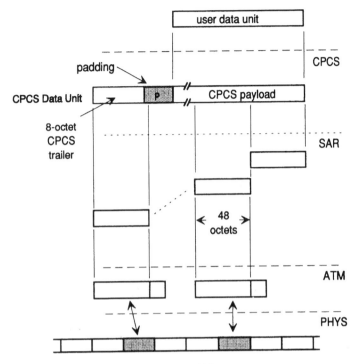

Figure 5.18 Data transfer using AAL5

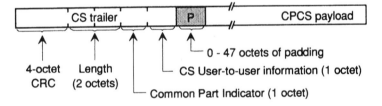

Figure 5.19 AAL5 CPCS data unit

'future-proofing': its use is not yet defined but it will probably be used for things like identifying management messages.

The length field specifies the number of octets of payload in the CPCS data unit, not including any padding. The CRC covers the entire CPCS data unit except for the CRC field itself. AAL5, with this 4-octet CRC in the CPCS data unit, gives a similar overall error performance to AAL3/4, with its 10-bit CRC in each SAR data unit.

The complete CPCS data unit is passed to the SAR sublayer which then segments it into a sequence of 48-octet SAR data units. The SAR sublayer does not add any overhead of its own, and it passes each SAR data unit to the ATM layer where it forms the complete payload of an ATM cell. Since the CPCS data unit is a multiple of 48 octets there will be no partially-filled cells.

The ATM-user-to-ATM-user bit in the ATM header, which is the least-significant bit of the payload type field (see Figure 5.10), is used to tell the remote AAL when the last segment of the CPCS data unit has been received. The ATM-user-to-ATM-user bit is carried unchanged through the network from the source ATM layer to the destination ATM Layer. It is set to 0 in cells containing the first or an intermediate SAR data unit, and to 1 in cells containing the final SAR data unit of a CPCS data unit. If a CPCS data unit fits entirely in a single cell the ATM-user-to-ATM-user bit will be set to 1. The use of the ATM-user-to-ATM-user bit avoids the need to add header or trailer bits to the SAR data unit.

At the receiving end the process is reversed to recreate the user data unit that was sent. The possible loss, misinsertion or corruption of cells by the network will be detected by the receiving CPCS by checking the CPCS data unit payload against the length field, by the the CRC check, or, if the last cell in the sequence is lost, by checking the received data unit against a predetermined maximum length or by reassembly buffer overflow.

So AAL5 achieves the same data transfer capability as AAL3/4, but in a simpler and more efficient way. In fact, though started later, AAL5 has actually overtaken AAL3/4 in its state of development and includes a few features that are not yet found in AAL3/4. The perceptive reader will have already spotted that there is no direct equivalent in AAL3/4 of the 1-octet CPCS user-to-user information transfer; though this can be achieved in other ways by AAL3/4, depending on the service specific convergence sublayers. But there are two other features of AAL5 that are not yet found in AAL3/4–AAL5 provides for the passing of congestion and priority information.

Using AAL5 the SSCS (or CPCS user if there is no SSCS) passes to the CPCS, in addition to the user data unit, an indication of the priority of that user data unit and an indication of whether it has encountered congestion. The CPCS passes these indications to the SAR sublayer with the CPCS data unit. The SAR sublayer then passes them to the ATM layer, which inserts them into the cell header. The priority indication is inserted into the cell loss priority bit (see Figure 5.10), indicating high priority (CLP = 0) or low priority (CLP = 1). The congestion indicator is inserted into the middle bit of the payload type field for user data cells. (The question of where congestion may have been encountered for data units that are about to be passed to the network may seem somewhat academic. But data units may have already traversed other networks (e.g. a Frame Relay network) before reaching the AAL).

It should be noted that the cell loss priority bit in any cell may have been changed by the source ATM layer or by the network if the traffic contract has been exceeded (if any of the cells carrying a user data unit are received with cell loss priority bit set to 1, indicating low priority, the corresponding priority indication passed to the user will also indicate low priority). The congestion indication bit in the ATM cell header may have been changed by the network if congestion has been encountered in transit. The priority and congestion indications passed to the remote user may therefore be different from those sent .

It is simply a question of time before these features are added to AAL3/4. The intention is that both AAL3/4 and AAL5 should support the same data tranfer capability, except for multiplexing which will remain the province of AAL3/4.

The traffic contract and policing (I.371)

The traffic contract for a connection specifies the characteristics of the traffic which the user will pass to the ATM network for delivery to the remote end, and the quality of service to be maintained on the connection (strictly speaking the traffic contract applies at the interface between the ATM and physical layers, the so-called physical layer service access point). Separate contracts apply to each direction of the connection.

A connection's traffic is characterised by a Peak Cell Rate (PCR) and a Cell Delay Variation (CDV) tolerance. The Peak Cell Rate identifies the maximum rate at which the user may send cells into the network. The CDV tolerance we will look at shortly. The network operates what is called Usage Parameter Control (UPC) to police the actual traffic sent into the network. If cells are passed to the network at a higher rate than the peak cell rate the network will either set the cell loss priority bit in the header of the offending cells to indicate low priority or discard the cells immediately. Setting the cell loss priority bit means that if the cell encounters congestion in the network it will be discarded in preference to cells that do not have the cell loss priority bit set. As a last resort the Usage parameter control may terminate the offending connection.

The user may himself use the cell loss priority bit to indicate cell priority. If he does then it will be necessary to specify two Peak Cell Rates for each direction of a connection, one relating to cells with CLP set to 0, the other relating to all cells (i.e. with CLP set to 1 and 0). Correspondingly, two different qualities of service would be agreed for the connection, specifying different objectives for cell loss, one relating to cells with CLP set to 0 and one relating to all cells.

On the face of it such an apparently simple parameter as Peak Cell Rate should be easy to deal with. The idea is simple: the Peak Cell Rate is just the maximum number of cells per second the user may pass into the network on a particular connection and still conform to the agreed contract. The same thing can be expressed in terms of the Peak Emission Interval (T), the inverse of the Peak Cell Rate, which is the minimum time permitted between conforming cells. If on a particular connection the network received any cells from the user within T seconds of the previous one they would be immediately identified as not complying with the traffic contract, and dealt with accordingly. For the reasons outlined below things are not quite this easy.

Consider for illustration an ATM connection supporting a Constant Bit Rate (CBR) circuit-emulation service. Since the bit stream input to the AAL (AAL type 1) by the user is clocked at a constant rate, we should expect cells to be generated at the regular rate of one every 47 octets (remember that AAL1 adds a one-octet sequence number for each cell leaving 47 for payload). We would therefore expect to see cells for this connection appear for transmission at perfectly regular intervals every T seconds, as shown in Figure 5.20. But the timing of the transmitted cell stream is independent of this and the cells have to wait for the next available slot to be transmitted. As the illustration shows,

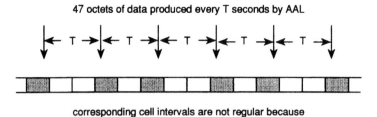

Figure 5.20   Cell delay variation caused by slot structure

this will cause spreading and bunching of the connection's cells as seen by the network, and some of the cells will appear within T seconds of the previous one for that connection. The network therefore thinks the traffic contract has been violated, but it has not.

There is more! This simple picture considers only cells arising from the CBR circuit-emulation service. In practice there will be management cells generated by the ATM layer and the physical layer. It is likely that the terminal will also have other connections in progress. So the cells of our CBR circuit-emulation service will be competing with numerous other cells for the available transmission slots, and will have to wait their turn in the transmission queue. This will cause even more spreading and bunching of the cells of our CBR circuit-emulation service, further increasing the likelihood that the network will think it has violated its traffic contract when in spirit it has not.

So the Peak Cell Rate on its own does not give us a reliable parameter against which to police compliance with the traffic contract. To overcome this problem a scheme has been devised in which the user specifies for both directions of each connection a second parameter, the CDV tolerance, in addition to the Peak Cell Rate. The CDV tolerance indicates the maximum deviation from the peak emission interval that may arise at the network interface from the distortions discussed above.

The scheme also specifies an algorithm by which each cell's conformance to the agreed peak cell rate can be gauged, including the agreed CDV tolerance. This algorithm, known as the Generic Cell Rate Algorithm (GCRA), in effect permits successive cells of a connection to be closer together than the Peak Emission Interval (by up to the CDV tolerance) so long as the average cell rate does not exceed the Peak Cell Rate.

Traffic contracts based on Peak Cell Rate characterisation and policing are effective and efficient for Constant Bit Rate services. But they can lead to poor utilisation of network resources for Variable Bit Rate services which often have mean cell rates that are much less than the peak rate. The ATM Forum's UNI specification includes traffic parameters based on mean rate characterisation for variable bit rate services. For this the user may, in addition to Peak Cell Rate and CDV tolerance, specify a Sustainable Cell Rate together with a Burst Tolerance parameter. It is likely that these parameters, or variations on the same theme, will be standardised by ITU-TS. For more detail on this the reader is referred to the ATM Forum specification (ATMF).

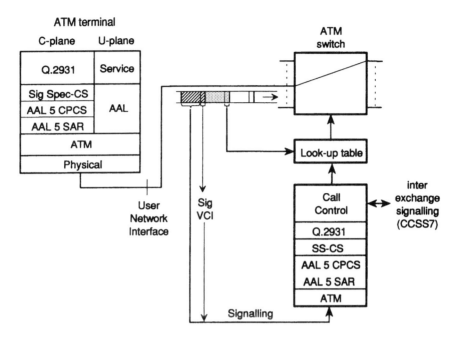

Figure 5.21 Signalling in an ATM environment

Signalling for ATM

The message is probably already clear to the reader that great things are expected of ATM and that it is seen by many as the fabric for broadband networking for many years to come. It promises unprecedented flexibility in terms of the variety of services it will support, including many that have not yet been dreamt of. We have seen something of how ATM can achieve this in the data path (the user-plane in standards parlance). But what about the signalling that will be needed to control this ever-increasing diversity of needs?

The ITU standards-makers believe it will take longer than the market will tolerate to develop the signalling protocols needed to exploit the full potential of ATM. A staged approach is therefore being taken with several 'releases' (Law 1994). Release 1 of ITU's B-ISDN User Network Interface signalling standard, Q.2931, is based closely on Q.931, the signalling standard used in ISDN. Figure 5.21 shows how this fits into the ATM scheme of things.

Viewed as an application, signalling looks like connection-oriented packet data and it should be no surprise that AAL 5 is used to support the transfer of signalling messages. A Signalling-Specific Convergence Sublayer has been defined, sitting above the common part convergence sublayer as shown. It is the job of the AAL and ATM layers to get the Q.2931 signalling messages to the call control process in the serving local switch as shown, the signalling using a pre-defined VCI.

There are few really significant differences between Q.2931 and Q.931 as used in the ISDN. For a readable account of Q.931 the companion volume

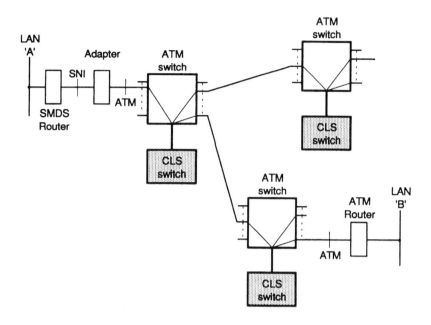

Figure 5.22   Connectionless service over ATM

ISDN Explained (Griffiths 1992) is recommended to the interested reader. However, the basic principle is as outlined in Chapter 3 when we looked at the ISDN origins of Frame Relay.

Future releases beyond Release 1 are likely to involve some substantial differences in approach, and readers are recommended not to hold their breath until they appear!

5.3   SMDS AND FRAME RELAY OVER ATM

It should be clear by now that ATM is innately connection-oriented. So how can it support connectionless services? The short answer is that it cannot. A connectionless service requires a network of connectionless switches; ATM switches are intrinsically circuit-oriented. What ATM can do is provide connections between the connectionless switches, and user access to these switches, as shown in Figure 5.22.

In effect the connectionless service is provided by a connectionless packet-switched network overlaid on the ATM network. The ATM network simply supplies PVPs between the connectionless switches and PVCs between the users and their serving connectionless switches. (In the literature the connectionless switches are usually referred to as connectionless servers, terminology that reflects the IT culture that has championed the connectionless cause rather than the telecommunications culture that is still strongly wedded to connections.)

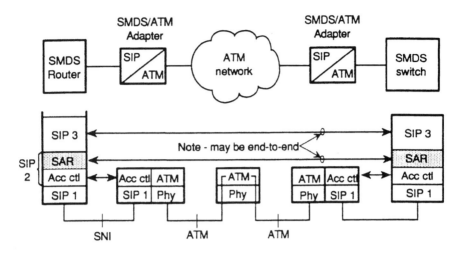

Figure 5.23 SMDS access over ATM

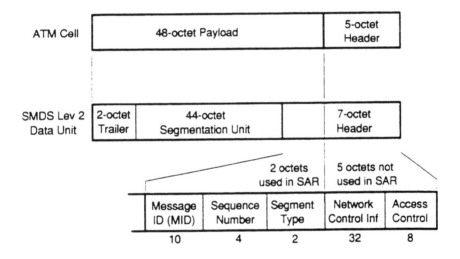

Figure 5.24 Mapping DQDB slots into ATM cells

SMDS over ATM

If we want to provide the SMDS service over an ATM network the simplest way would be to use SMDS/ATM adapters, as shown in Figure 5.23 to convert between the SMDS interface protocol and ATM, at the router and at the serving SMDS switch, and connect them by an ATM PVC. Note that the SMDS switches forming the connectionless overlay network would also be interconnected by ATM connections.

Because (by design) DQDB uses a 53-octet slot and ATM uses a 53-octet cell the functionality of the SDMS/ATM adapter can be very simple. Figure 5.24

shows that all the information needed in the SMDS level 2 data unit to reassemble the SMDS packet actually fits into the 48-octet payload of a single ATM cell (see also section 4.2 and Figure 4.10).

All the SMDS/ATM adapter needs to do is take the least significant 48 octets of the SIP level 2 data unit and pass it to the ATM layer as the payload of an ATM cell. The ATM layer then inserts the appropriate VPI/VCI and other control fields into the cell header and passes the cell to the physical layer for transmission over the PVC. The adapter at the SMDS switch would reverse this process to recover the SMDS level 2 data unit. The 5 octets of SDMS level 2 data unit header that are not sent over the ATM connection (the access control field and the network control information) can be recreated by the remote Adapter, since they are always set to fixed bit patterns.

So the Adapter is really only concerned with DQDB bus access control and mapping the 48 octets of payload between the ATM cell and the SMDS level 2 data unit. It does not get involved in segmentation and reassembly of SMDS packets. The SAR function is implemented in the SMDS router and the SMDS switch, as shown in Figure 5.23.

There is in fact no need for the SMDS switch to reassemble the SDMS packets before routing them out. Since all the information needed by the switch to validate, police and route the packet is contained in the first segment of the packet (the BOM (or SSM) segment) most switch designs route the packet over the SMDS network as segments rather than as a fully assembled packet. This keeps switching delays to a minimum because the switch does not have to wait for the complete packet to be received before starts sending it on. So in practice segmentation and reassembly usually operates end-to-end between the SMDS CPE. This would always be the case where ATM connections are used to interconnect the SMDS switches, as here.

The adapter functionality at the SMDS switch end may be integrated with the switch itself, as shown in Figure 5.25, or in the SMDS router.

The alert reader will have noticed that the segmentation and reassembly process used by SMDS level 2 is the same as that used in ATM's AAL3/4. Again, this is not an accident, and it reflects the high degree of co-operation between the standards-makers involved in SMDS and ATM. Indeed, the co-operation goes further. As part of the B-ISDN, ITU-TS has defined a broadband connectionless data service (I.364) that is basically the same as SMDS, including packet formats. It is based on an infrastructure of connectionless switches interconnected by ATM connections, as illustrated above.

Just to add to the confusion of names it is worth noting that ETSI's version of SMDS is known as the Connectionless Broadband Data Service, or CBDS (ETS 300 217).

Figure 5.26 shows how the SMDS protocol stack relates to that of the broadband connectionless data service (Yamazaki, Wakahara and Ikeda 1993). It can be seen that SIP level 2 corresponds to the ATM layer together with the SAR sublayer of AAL 3/4, and that SIP level 3 corresponds to the common part of the AAL 3/4 convergence sublayer (CPCS) plus what is called the connectionless network access protocol (CLNAP).

The CLNAP basically takes the user's data block (typically a LAN packet),

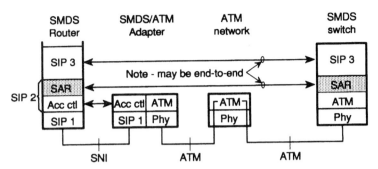

Figure 5.25   Putting the adapter into the SMDS switch

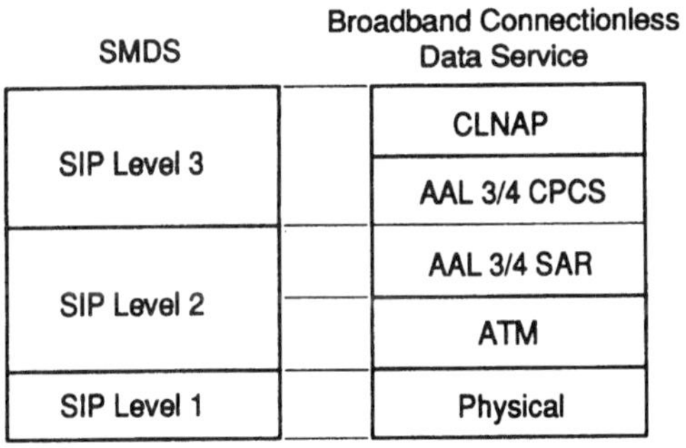

SMDS		Broadband Connectionless Data Service
SIP Level 3		CLNAP
		AAL 3/4 CPCS
SIP Level 2		AAL 3/4 SAR
		ATM
SIP Level 1		Physical

Figure 5.26   SMDS and BISDN connectionless data services

adds the E.164 source and destination addresses and other control information such as higher-layer protocol identifier (to tell the recipient how to process the received data block), quality of service indication and so on, and passes the resulting data unit to the AAL 3/4 layer. When the AAL 3/4 CPCS has added its header and trailer to this the resulting data unit has the same format as the SMDS packet. Segmentation and reassembly are then the same for both SIP and CBDS in terms of creating the 48-octet payload for the SIP slot and the ATM cell.

It will be some time before the broadband connectionless data service arrives. But it is clear that future interworking with and migration from SMDS will be comparatively easy. Indeed, we can see by comparing the protocol stack in the SMDS switch in Figure 5.25 with that for the broadband connectionless data service in Figure 5.26 that we are already nearly there. So investment in SMDS now will be good for a long time to come.

Frame Relay over ATM

As in the case of SMDS, an adapter is needed to interface a native-mode Frame Relay terminal to an ATM network, as shown in Figure 5.27. But remember that Frame Relay is a service rather than a technology and it can also be provided on an ATM terminal if the appropriate ATM adaptation

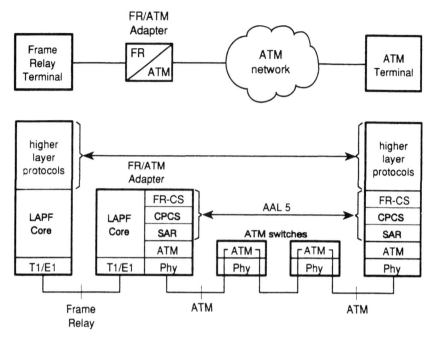

Figure 5.27 Frame Relay over ATM

function is implemented.

As a connection-oriented data service Frame Relay falls into service class C and could use either AAL 3/4 or AAL 5. However, the rigour of AAL 3/4 is not really consistent with Frame Relay's lightweight data transfer protocol and a Frame Relay Service Specific Convergence Sublayer has been defined for AAL 5 (I.365.1), shown as FR-CS in Figure 5.27. This implements Frame Relay's LAPF core protocol so that the service seen by the AAL user, at the ATM terminal on the right of Figure 5.27, is the same as for the native-mode Frame Relay terminal shown on the left. Neither user is directly aware of the ATM infrastructure.

The FR/ATM adapter terminates the LAPF core protocol to the Frame Relay terminal on the left in Figure 5.27, and terminates the ATM protocol into the ATM network. The user information field of the Frame Relay frame is passed between the two halves of the adapter in each direction. Indications of congestion are carried differently by Frame Relay (which uses the FECN and BECN bits) and ATM (which uses the middle bit of the payload type field), and congestion indications may also be passed across (I.555).

There are three interworking situations that can arise in carrying the Frame Relay service over an ATM network. With a little imagination the reader can visualise all three cases in Figure 5.27. Firstly, both terminals may have Frame Relay network interfaces, in which case both would need an Adapter. Secondly, both terminals may have ATM network interfaces, in which case neither would need an adapter. Thirdly, where one terminal has a Frame Relay network interface and the other an ATM network interface, which is the case illustrated directly in Figure 5.27, only the one with the Frame Relay network interface needs an adapter.

Note that in the general case there may be a Frame Relay network between the Frame Relay terminal and the adapter; the adapter would then be an internetworking unit (this would lead to minor differences in adapter functionality since, if the adapter were shared by a number of Frame Relay users some sort of registration procedure would be needed). But this does not change the basic principle.

The natural migration path is that over time (perhaps a decade or more) native-mode Frame Relay terminals will be displaced by their ATM equivalents, just as dedicated Frame Relay networks will be displaced by the increasing penetration of ATM networks. The real point to be understood here is that Frame Relay is a bearer service and will survive the transition to ATM.

5.4   ATM IN LOCAL AREA NETWORKS

Local area networking really began in the late 1970s, and the following decade saw extremely rapid growth fuelled to a large extent by the LAN standards published by the IEEE: 802.3 for the Ethernet type CSMA/CD network, 802.4 for the Token Bus network, and 802.5 for the Token Ring network. These LANs all use a shared-medium topology.

At the time of their introduction LANs were regarded as broadband. They offered a shared-medium channel running at megabit speeds, and most applications at the outset were straightforward data or text that did not make great demands on bit rate. But during the 1980s developments were very rapid in personal computers, and the applications have become much more bandwidth-hungry in their communication requirements, both in terms of the local on-net traffic and the inter-LAN traffic, a trend that is now beginning to accelerate.

What has happened is that the shared medium that first gave personal computers a window on a wider world has now become a bottleneck. Where previously an Ethernet supported perhaps 100 stations, it may now be divided by bridges into two or more segments to increase the LAN capacity available to each station. As a natural continuation of this we have seen the beginning of a migration away from shared-medium to star topologies, as in switched Ethernet, for example, which makes 10 Mbit/s (or even 100 Mbit/s) of capacity available to every station.

At the heart of this development are intelligent switching hubs, as illustrated in Figure 5.28, which can support a wide range of network types, allowing a company to flexibly mix and match its LANs and servers to reflect changes in the organisation and to embrace new technology. ATM is one of these new technologies and hubs based on ATM switches are now available that support both the interconnection of legacy LANs (that is, the old-fashioned but paid-for pre-ATM equipment) and ATM stations.

Legacy systems will be with us for a long time, but the long-term direction is now clear. During the next decade we will see the growth of local area networking based on switched star networks. Though there are a number of established technologies, such as 10 Mbit/s Ethernet, that will be used for this

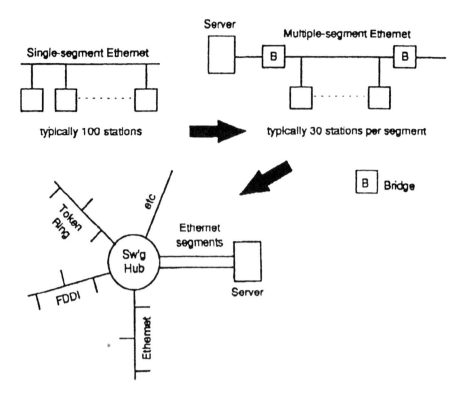

Figure 5.28 LAN evolution to switching hubs

both because they exploit existing investment and because they will satisfy many requirements more cheaply, there is no doubt that the overall trend will be towards ATM, and much of the present momentum in ATM is in developments for local area networking. The ATM Forum in particular has for some time been developing early intercepts of the emerging international standards (ATMF), and the equipment needed to build ATM local area networks of the sort illustrated in Figure 5.29 is commercially available now based on the ATM Forum specifications. In fact at the time of writing the only thing that is missing is the ATM public wide area network—but it is coming!

One of the drivers for this is multimedia. ATM intrinsically supports communication for multimedia services and has the flexibility to cater for the new requirements as they develop. The other big driver is ATM's ability to bridge the gap (or chasm) between the local and the wide area. As LANs become ever more deeply embedded as part of a company's corporate infrastructure, the demands of wide area LAN interconnection will continue to increase. The use of ATM wide area networks for this promises a seamlessness in communications that has hitherto been notable by its absence. Indeed, by removing the barrier of distance in this way ATM is the key to Total Area Networking.

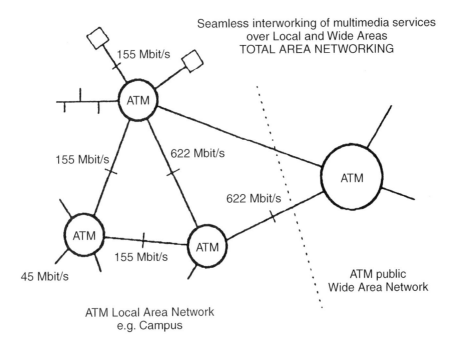

Figure 5.29 ATM in the local area

5.5 SUMMARY

In this chapter we have looked at the basics of ATM and how it works. In effect there are two distinct aspects. On the one hand ATM provides a universal switching fabric supporting the ATM bearer service. On the other hand the services to be carried have to be adapted so that they fit on to this ATM bearer service. The bearer service can be tailored to meet the requirements of the service being carried, and we have looked at how the network polices each connection to make sure that the appropriate quality of service is maintained. We have also looked at how services are adapted to map on to the bearer service.

There is often confusion about how Frame Relay, SMDS and ATM relate to each other. The truth is that Frame Relay and SMDS are services whilst ATM is a network infrastructure that can carry them. So the introduction of ATM networking will not make Frame Relay or SMDS obsolete any more than it will make any other service, such as telephony, obsolete. We have looked at how Frame Relay and SMDS are carried over an ATM network. In particular we have seen how connectionless services such as SMDS need connectionless switches overlaid on the connection-oriented ATM network.

ATM was originally conceived and developed by CCITT (now ITU-TS) as the basis for B-ISDN, the wide area network that would be all-things-to-all-men. But it would not wait for the long gestation of international standardisation before being delivered to the world. Its flexibility and richness of features quickly established it as the infant technology for the LANs of the future. The increasing demand for ever higher bit rates by LAN applications has effectively sealed the fate of shared-medium LAN technologies and is driving a move to switched-star LANs. ATM is at the forefront of this and LANs based on ATM switches represent the shape of things to come.

ATM will therefore provide the universal switching fabric for both local and wide area networking, enabling seamless interworking between them, taking the distance out of communication. ATM is a therefore a key milestone on the road to total area networking.

REFERENCES

General

Griffiths, J. M. (1992) *ISDN Explained.* John Wiley & Sons.

Law, B. (1994) Signalling in the ATM network. *BT Technology Journal,* **12**, No. 3.

Yamazaki, K., Wakahara, Y. and Ikeda, Y. (1993) Networks and switching for B-ISDN connectionless communications. *IEICE Transactions on Communications*, **E76-B**, March.

Standards (note that the following lists are not exhaustive)

ITU-TS
I.121 Broadband Aspects of ISDN
I.150 BISDN ATM Functional Characteristics
I.211 BISDN Service Aspects
I.311 BISDN General Network Aspects
I.321 BISDN Protocol Reference Model and its Application
I.361 BISDN ATM Layer Specification
I.362 ATM Adaptation Layer (AAL) Functional Description
I.363 BISDN ATM Adaptation Layer (AAL) Specification
I.364 Support of broadband connectionless data service on BISDN
I.365.1 Frame Relay Service Specific Convergence Sublayer
I.371 Traffic and Congestion Control in BISDN
I.555 Frame Relaying Bearer Service Interworking
Q.931 ISDN user–network interface layer 3 specification for basic call control
Q.2931

ATM Forum

ATM User–Network Interface Specification

ETSI

ETS 300 217: Connectionless broadband data service

6

A Telecommunications View of the Total Area Network

Intelligence is quickness to apprehend as distinct from ability,
which is capacity to act wisely on the thing apprehended

Alfred North Whitehead

We have seen in Chapter 3 that the Integrated Digital Network (IDN) and ISDN evolved from the analogue Public Switched Telephone Network (PSTN). And we have seen as part of this evolution how the network has been logically segregated into a 'switched information' subnet (the user-plane or U-plane in ISDN parlance) and a 'signalling' subnet (the control-plane or C-plane). In terms of Whitehead's dictum we can associate intelligence with the C-plane and ability with the U-plane. This separation of switching and signalling arises naturally from the essentially different nature of the technologies used for switching and signalling. As we will see, the two planes can evolve separately to exploit advances in their respective techniques and technologies.

So far this book has focused mainly on the evolution of the user-plane—from analogue voice, through 64 kbit/s circuit switching, Frame Relay (as an ISDN bearer service), and in due course ATM. Each of these switching techniques and technologies provides additional flexibility in the range of services that can be offered to the user, and in the way that distance is perceived, or preferably not perceived. Remember, the overall aim in telecommunications is to take the distance out of information—that is, Total Area Networking. In this chapter we will look at the part the control-plane plays in Total Area Networking, and how it is evolving.

One of the increasing important factors shaping developments in telecommunications networks and services is competition. The long-heralded liberalisation of telecommunications is now well under way almost everywhere, the traditional monopoly suppliers, the national PTOs, being forced to share their market with newcomers, the 'Other Licenced Operators' or OLOs. In this environment the customer is 'king', and if one service provider does not meet his needs another will. Those service providers will therefore prosper who can respond most quickly to new customer demands. We will see that the control-plane holds the key to such rapid response, and that it exercises this power by virtue of its intelligence.

Competition is, of course, not confined to telecommunications. The globalisation of business and commerce that modern telecommunications has done so much to facilitate is itself bringing new opportunities to gain competitive advantage: indeed, that is its justification. But the pace of change is rapid and competitive advantage can quickly change hands as competitors play leap-frog in their search for success. It is clear that the most successful companies will be those with the 'agility' to respond quickly to their competitors' activities and to the developing expectations of the customer. Compared with traditional private networks, which can quickly be overtaken by advances in technology and which tend to divert a company's resources away from its core business, VPNs can make an important contribution to a company's agility. As the platform for the implementation of VPNs therefore, the Intelligent Network may be expected to become an increasingly important part of every major company's service infrastructure as public network IN capabilities develop. In effect, VPNs will be a major step on the road to Total Area Networking.

This chapter is about the evolution of the IDN/ISDN to become the Intelligent Network or IN.

6.1   SIGNALLING IN THE NETWORK—CCSS7

Before embarking on this story we will lay the foundations. Since this chapter focuses on the C-plane we must begin with a brief review of signalling, the language of the C-plane. In Chapter 3 we looked briefly at ISDN signalling between the user and the ISDN network and how a simple call would be set up and cleared (Figures 3.1 and 3.2). Here we extend this to include signalling between the switches which uses a similar message-based signalling protocol known as CCITT Common Channel Signalling System Number 7, or CCSS7 (or even just C7) for short.

In its full glory C7 is a very comprehensive and necessarily complex protocol and justifies a book to itself much bigger than this one! We will limit ourselves here to the essentials needed to develop our story. Figure 6.1 shows an example of the signalling involved in setting up and clearing a basic call, assuming ISDN terminals at both ends (see also Figures 3.1 and 3.2).

The calling terminal, a digital telephone say, initiates call set-up by sending

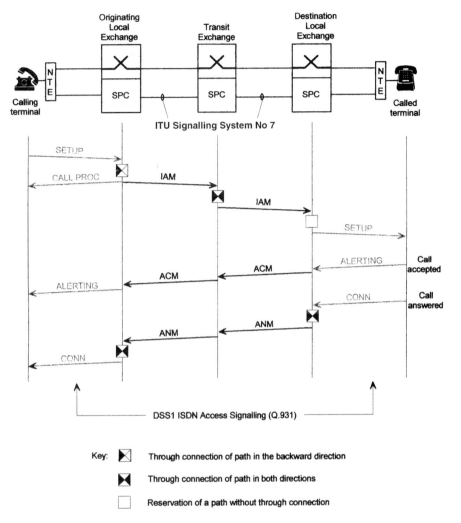

Figure 6.1 Signalling for basic call set-up in the ISDN (Figure 6.4(a))

an ISDN SETUP message to the originating Local Exchange. This SETUP message contains the calling and called party numbers and other information needed to establish an appropriate connection (such as whether a digital connection is needed from end-to-end or whether a partly analogue connection would do). The originating Local Exchange acknowledges receipt of this message by returning an ISDN CALL PROCEEDING message indicating that the network is attempting to set the call up.

The Call Control process in the originating Local Exchange then translates the ISDN SETUP message into a corresponding CCSS7 message, which is an Initial Address Message or IAM. This Initial Address Message is routed through the signalling subnet until it reaches the Local Exchange serving the called party (the destination Local Exchange), the routing decision at each switch *en route* being based on the called party's number and any other pertinent information contained in the IAM (such as whether a satellite link is acceptable).

The destination Local Exchange translates the Initial Address Message into a corresponding ISDN SETUP message which it delivers to the called party. The called party accepts the call by returning an ISDN ALERTING message to the destination Local Exchange. The ALERTING message is translated into a CCSS7 Address Complete Message (ACM) which is passed back to the calling terminal as an ISDN ALERTING message as shown. The ACM both indicates to the other exchanges involved in the connection that the destinationLocal Exchange has received enough address information to complete the call and passes the alerting indication (i.e. that the called party is being alerted) to the originating Local Exchange.

The speech path is shown as switched through in the backward direction at the originating Local Exchange on receipt of the SETUP message and switched through in both directions at the Transit Exchange on receipt of the IAM. This allows the caller to hear any 'in-band' signalling tones sent by the network (for a variety of reasons, not all call attempts succeed).

The called telephone now rings and the originating Local Exchange sends ringing tone to the caller.

When the call is answered (i.e. the handset is lifted) the called telephone generates and sends an ISDN CONNECT message to the destination Local Exchange, which it translates into the corresponding CCSS7 Answer message (ANM). This is passed back to the calling Local Exchange where it is translated back into an ISDN CONNECT message and passed to the calling terminal. At each switch *en route* any open switch points are operated to complete the connection in both directions, giving an end-to-end connection, and the call enters the 'conversation' phase. Billing for the call usually starts at this point.

Note that in the case of ISDN access there is a distinction between accepting the call and answering it. The reason for this is that, unlike a PSTN access, the Basic Rate ISDN customer interface takes the form of a passive bus that can support simultaneously a number of different terminals (up to eight), of different types (such as fax machines, telephones, personal computers, and so on). The destination Local Exchange does not know until it receives the ALERTING message from the called party whether he has an appropriate terminal connected to the interface that can take the call (amongst other things the SETUP message may carry compatibility information that the terminals may use to ensure compatibility between calling and called terminals). If there were not an appropriate terminal connected to the called access the call would not be accepted.

At some later time the calling party (say) clears the call. This is signalled to the originating LE by means of an ISDN DISCONNECT message, as shown in Figure 6.2. The originating Local Exchange then initiates release of the ISDN access circuit by returning an ISDN RELEASE message, acknowledged on completion by the calling terminal sending a ISDN RELEASE COMPLETE messaage. Release of the inter-exchange circuit is signalled to the Transit Exchange by a CCSS7 Release (REL) message, completion of which is

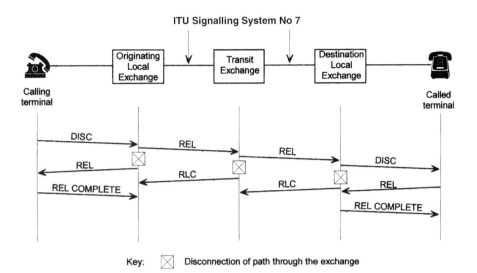

Figure 6.2 Normal call clear sequence using CCSS7

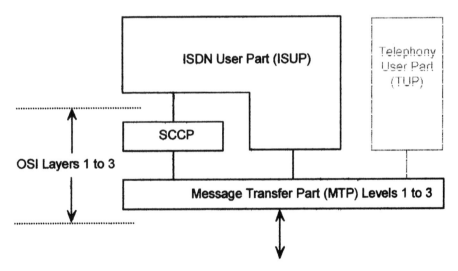

Figure 6.3 The CCSS7 protocol stack for ISUP

signalled back by a CCSS7 Release Complete (RLC) message. Successive circuit segments are released in a similar way as shown. A similar process in the other direction is used if the call is cleared by the called party (though there is also the option for the called party to suspend the call for a short time by replacing the handset before resuming the call).

The basic CCSS7 protocol stack is shown in Figure 6.3. It is a layered protocol, but was defined before the publication of the OSI Reference Model (RM), and the CCSS7 levels of protocol, though similar, do not correspond exactly with the OSI layers. The alignment of CCSS7 with the OSI Reference Model is a comparatively recent development as described below.

The Message Transfer Part, or MTP, provides for the reliable, error-free transmission of signalling message from one point in the CCSS7 signalling subnet (referred to as a Signalling Point) to another such point. It is itself organised as three distinct functional levels—similar to but not the same as the lowest three OSI layers.

MTP Level 1—the physical level—is usually referred to as the Signalling Data Link. It provides a physical transmission path (usually a 64 kbit/s time-slot in a higher-order multiplex) between adjacent Signalling Points.

MTP Level 2, usually known as Signalling Link Control, deals with the formation and sending of Message Signal Units (MSUs) over the Signalling Data Link, checking for errors in transmission using a cyclic Redundancy Code added to the MSU before transmission (in effect a form of parity check), and correcting any such errors by retransmitting the MSU. In this way MTP level 2 ensures that signalling messages get neither lost nor duplicated. It also operates a flow control procedure for message units passed over the signalling link. Like level 1, level 2 operates only between adjacent Signalling Points. So a signalling 'connection' between an originating and destination Local Exchange involved in setting up a call actually involves a number of independent signalling links in tandem.

MTP Level 3 is concerned with routing signalling messages to the appropriate point in the CCSS7 signalling subnet based on unique 14-bit addresses, known as Signalling Point Codes, assigned to each such point in the signalling subnet. Routing is predetermined with alternative routes specified for use if the primary route becomes unavailable. So at each Signalling Point reached by a signalling message a decision is made as to whether the message is addressed to that Signalling Point or is to be routed onward to another. When used to route signalling messages in this way a Signalling Point is operating as a Signalling Transfer Point or STP.

The ISDN User Part, or ISUP, uses the services provided by the MTP. It is concerned with the procedures needed to provide ISDN switched services and embraces the functions, format, content and sequence of the signalling messages passed between the signalling points. An example of ISUP at work is shown in Figure 6.1.

Whilst the focus here is on the ISDN it should be realised that the first version of CCSS7, published in 1980, did not cover ISDN services, which were not defined until 1984. The 1980 CCSS7 standard defined the Telephony User Part, or TUP, which does for analogue telephone services what ISUP does for ISDN services. In practice the two ISDN and Telephony User Parts will co-exist, perhaps for many years, before TUP is entirely supplanted by ISUP. But for clarity and brevity here, and because we are looking to the future, we focus on ISUP.

One of the shortcomings of the 1980 version of CCSS7 was that signalling was defined in terms of the messages that passed between adjacent exchanges. This was fine for analogue telephony services. But the ISDN, with its powerful signalling between user and network, brought a much wider range of services into prospect. Many of these services require signalling messages to be passed between the originating and destination Local Exchanges without the intervention of intervening exchanges *en route*. Indeed, in some cases signalling is required between the Local Exchanges even in the absence of a connection being established between them.

This requirement found the MTP wanting and in 1984 the Signalling Connection Control Part, or SCCP, was added to CCSS7 to provide greater flexibility in signalling message routing. Whilst the Telephony User Part (TUP) uses only the services of the MTP, the ISDN User Part (ISUP) also makes use of the SCCP as shown in Figure 6.3. The SCCP was designed to provide the (by then) standard OSI network layer service, supporting both connectionless and connection-oriented methods of message transfer. In effect, it created a packet-switched network within the signalling subnet by means of which any Signalling Points can send signalling messages to any other Signalling Point, independent of switched connection in the switched information subnet. We will see below that even this is not the complete picture for ISUP, but we break the story here in order to renew it later when we have introduced the idea of the Intelligent Network.

6.2 THE TRANSITION TO THE INTELLIGENT NETWORK

In principle the IDN and ISDN are sufficiently flexible to provide services tailored to each company's specific requirements. Providing such customised services means making (part of) the public network behave as though it were the company's own private network, i.e. a Virtual Private Network. In practice, however, this flexibility has not been achieved with the IDN/ISDN. The potential flexibility of stored program control—that is, software control of switches—has not been realised because of the way the call control software and its associated data has been implemented in the exchanges.

The problem stems from the fact that the service information relating to a customer's lines is stored in the serving Local Exchange. the companies with the greatest needs—those with the most to gain from customised services—are large and spread over many sites, indeed often over a number of countries. So the service information relating to such companies is distributed over a potentially large number of Local Exchanges, perhaps hundreds. Indeed, when looking at the collective requirements of corporate customers the information is distributed over all Local Exchanges, perhaps thousands. The problem of managing such a large distributed database and the associated co-ordination of customised call control has provided prohibitive.

The solution to this co-ordination problem has been to separate the 'advanced' service logic and the associated customer information from the 'basic' call control logic and switches. Basic call control continues to reside in the Local Exchange. But the advanced service logic defining the customer's requirements is centralised in what is an intelligent database as shown in Figure 6.4. Adding this centralised network intelligence to the IDN/ISDN creates what has become known as the Intelligent Network, or IN. As we will see, with this arrangement it becomes comparatively straightforward to

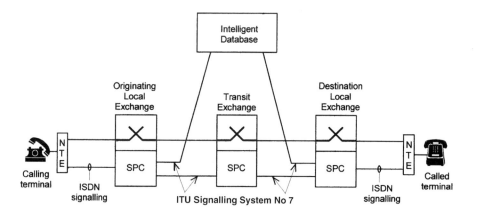

Figure 6.4   The IDN/ISDN + centralised network intelligence = IN

manage a comprehensive, up-to-date picture of a corporate customer's 'private' network requirements and to co-ordinate switching operations throughout the network in order to implement these requirements. CCSS7 continues to be the signalling system of choice for IN operations.

6.3   IN ARCHITECTURE AND TERMINOLOGY

The main building blocks of the IN are the Service Switching Point (SSP) and the Service Control Point (SCP) as shown in Figure 6.5. The SSP is (usually) part of the Local Exchange whose call control software has been restructured to separate basic call control from the more advanced call control needed for Intelligent Network Services (this terminology is somewhat circular—Intelligent Network Services are simply those services which need the Intelligent Network capability).

Basic call control looks after the basic switching operations that take place in an exchange. It has been restructured to incorporate what are known as Points In Call (PICs) and Detection Points (DPs) as defined points in the basic call control state machine. At these points trigger events may be detected and call processing temporarily suspended whilst reference is made to the centralised Service Control Point (SCP) to find out how the call should be handled from that point. Typical trigger events include such things as recognition of the Calling Line Identity (CLI) and recognition of dialled digit strings.

The Service Control Point (SCP) is a general-purpose computing platform on which the advanced service logic needed for Intelligent Network Services is implemented together with the information that defines each corporate customer's network services. It must be fast to provide the rapid response needed and to handle the potentially very high traffic levels arising from its central location. And of course it has to be reliable. To meet these stringent

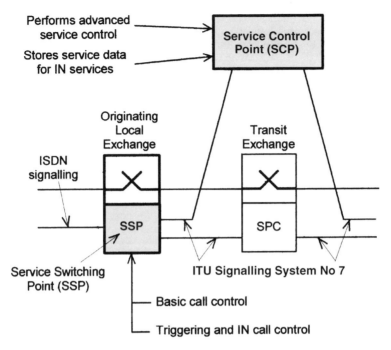

Figure 6.5 IN architecture and terminology

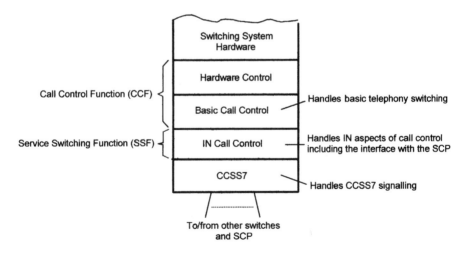

Figure 6.6 The Service Switching Point (SSP)

requirements more than one SCP is normally provided. In practice there may be a dozen or more.

Figure 6.6 introduces more jargon. The Service Switching Point software within the exchange consists of a Call Control Function (CCF) and a Service Switching Function (SSF). The Call Control Function looks after the basic call control needed for simple telephony switching operations. The Service

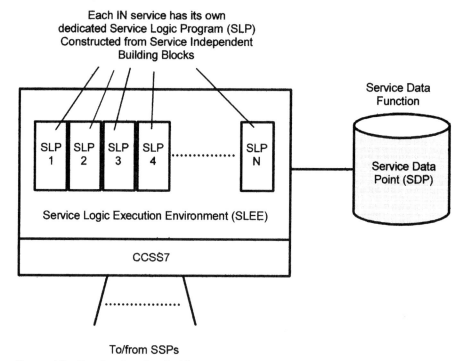

Figure 6.7 The Service Control Point (SCP)

Switching Function provides the control itnerface with the Service Control Point (and with another IN network element known as the Intelligent Peripheral (IP) that we will look at shortly).

And there is yet more jargon! The Service Control Point (SCP) contains the advanced service logic needed to implement Intelligent Network Services, as shown in Figure 6.7. Each such service, such as 0800 Freefone (which we will look at in more detail below), requires a Service Logic Programme (SLP) which is built from Service Independent Building-blocks (SIBs) together with the service information defining the corporate customer's detailed requirements which is held in the associated Service Data Point (SDP). Service Independent Building-blocks would typically include such operations as numer translation, connecting announcements, charging, and so on. Strictly speaking the Service Data Point need not be co-located with the Service Control Point. But it usually is and we will assume here that the Service Data Point resides within the Service Control Point.

The Service Logic Execution Environment (SLEE) is the generic software that controls the execution of the Service Logic Programmes. It interworks with the basic call control process and the simple switching functions in the Service Switching Point and screens the Service Logic Programmes from the low-level SCP–SSP interactions and controls the impact of new Service Logic Programmes on existing IN Services.

CCSS7 needed to be extended to support IN Services. In particular a Transaction Capabilities (TC) Application Part has been added to support

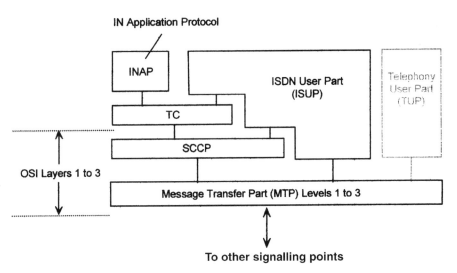

Figure 6.8 The CCSS7 protocol stack for IN

signalling that is not related to switched connections, as shown in Figure 6.8. And here we can complete the picture for ISUP. Since some ISDN supplementary services involve signalling that is not related to switched connections, ISUP may also use the services of the Transaction Capabilities Application Part as shown.

Examples of non-connection-related signalling include Operation Administration & Maintenance (OA&M) messages, customer-to-customer data transfer (via the signalling subnet), IN applications such as signalling between the SSP and SCP, and signalling for cellular mobile telephone networks (where roaming may be thought of as a particular example of an IN service tailored to a specific situation).

A new protocol has been developed specifically for the Intelligent Network, the so-called Intelligent Network Application Part (INAP), which may be considered part of the CCSS7 protocol suite. Looking upwards, INAP interfaces directly with the Service Logic Execution Environment of the SCP. The Transaction Capabilities (TC) Application Part supports TC users such as INAP (and MAP, the Mobile Application Part). It provides the OSI Session layer service together with dialogue control and is responsible for managing communications with remote TC users.

INAP defines the CCSS7 signalling messages relating to IN services and the functions and interactions they cause (in the form of finite state machines). INAP is in turn defined in terms of Abstract Syntax Notation 1 (ASN.1) making it independent of the computing platform and porotable to any processing environment. This is an important consideration in the quest for IN products that will actually interwork and in providing network operators with a means of enhancing their systems 'in-house' rather than being continually dependent on the manufacturers. INAP, like SCCP, is closely aligned with OSI standards. It is based on the OSI Remote Operations Service Element (ROSE).

It is worth noting here that the Signalling Connection Control Part (SCCP) can ensure that IN messages destined for a failed SCP are automatically re-routed to an operational one.

6.4   EXAMPLES OF IN SERVICES

0800 Freefone

Probably the best known example of an IN Service is 0800 Freefone, which we will use here to illustrate the main ideas of the Intelligent Network and to introduce another IN network element, the Intelligent Peripheral (IP) mentioned above. The Freefone example is illustrated here with reference to a hypothetical case of a large insurance company with branches in high streets up and down the country, six area offices each dealing with the administration of the high street branches within their respective geographic areas, and a national headquarters office in the capital.

A typical requirement of such a company, illustrated for clarity in Figure 6.9, would be to have a single, unique telephone number covering the whole country, such that:

- during normal office hours calls to that number would be routed to the high street branch nearest to the caller;

- out of normal office hours calls are routed to the area office covering the caller's location;

- when the area offices are closed calls should be routed to the headquarters office where they would be handled by the company's automated call handling system;

- the calls should be free (to the caller).

This service requirement can be satisfied by using the 8000 Freefone service whereby the company has an easily remembered number, say 0800 123abc. The branch and area offices will naturally have a variety of unco-ordinated telephone numbers. the list of translations from 0800 1234abcd to the appropriate branch or area office telephone number is stored in the central IN database, i.e. the SCP, together with the time of day, day of week, and day of year routing schedule.

It is 09:31 on a normal Monday morning and a customer (or potential customer) of the company dials the company's national number, 0800 123abc. Though it is not necessary, we will assume in what follows that both caller and company are ISDN-based, so ISDN access signalling (Q.931) is used at both ends of the call as shown in Figure 6.10. The caller's telephone number is 01234 567abc.

The basic call control process in the serving SSP, by reference to its Trigger

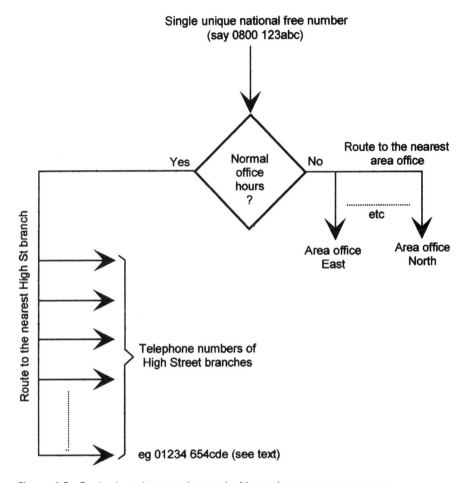

Figure 6.9 Typical service requirement of large insurance company

Table, recognises the 0800 code as involving an IN service. Basic call control is then suspended and the SSP sends a CCSS7 signalling message to its SCP giving the dialled number, 0800 123abc, and the caller's telephone number, 01234 567abc (the CLI). By reference to its routing schedule for that particular 0800 number (123abc) the SCP knows that for that time of day, day of week, and day of year the call should be routed to the high street branch nearest to the caller. And by reference to the CLI the SCP knows that the telephone number of the nearest such branch office is 01234 654cde (the caller and his nearest branch office may not have the same area code; it depends on the geography).

The SCP then returns a CCSS7 signalling message to the SSP advising that the actual destination number for the call is 01234 654cde and that the aller should not be charged for the call. On receipt of this message the SSP resumes basic call processing, the call is routed through the network to 01234 654cde in the usual way, and the call is charged to the insurance company.

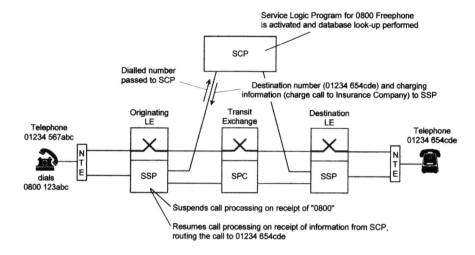

Figure 6.10   0800 Freefone—an IN service

0800 Freefone with user interaction

Let us suppose now that the insurance company takes over a financial services company and wants to incorporate the associated investment business into the existing company structure and processes. A new requirement is to direct customers' calls to the right sales team, for either insurance or investment business. It is important to keep both types of business strictly separate because they come under different regulators. We need to get the caller to indicate the nature of his enquiry. At the appropriate point in the call we get the IN to send him an announcement saying 'If you want investment services please enter 2. For insurance services please enter 3'.

To do this we need to modify the Service Logic Programme in the SCP to reflect the new requirement. And we need an additional network element, the Intelligent Peripheral (IP) as shown in Figure 6.11, to provide the customised announcements to the caller and detect any MF digits he or she dials.

The Intelligent Peripheral provides 'specialised resources' such as customised announcements, concatenated announcements, MF digit collection, signalling tones, and audio conference bridges. In due course, no doubt, it will also incorporate capabilities such as voice recognition to simplify the user's interface with the network. It is connected by traffic circuits to the switch so that it can be connected to the right user at the right time. And it has CCSS7 signalling links to the SSP and SCP. There is a design choice of controlling the Intelligent Peripheral directly from the SCP or indirectly via the SSP. Clearly, with such comprehensive capabilities the Intelligent Peripheral is going to be an expensive piece of kit, and it may serve more than one SSP depending on cost/performance considerations.

Figure 6.12 shows how the Service Logic Programme would be modified to meet the requirement for customer interaction. Again, it is 09:31 on a normal

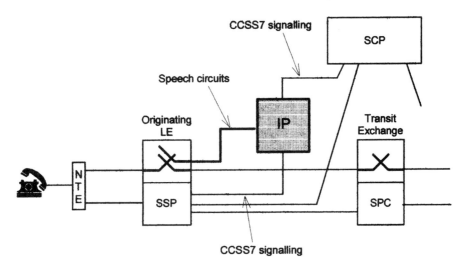

Figure 6.11 The Intelligent Peripheral (IP)

Tuesday morning, and the caller dials 0800 123abc, the national number for the insurance/financial services company. The call set-up proceeds as before up to the point at which the SSP suspends basic call processing and sends the CCSS7 signalling message to the SCP containing the dialled number, 0800 123abc, and the caller's number, 01234 567abc. But this time, the Service Logic Programme in the SCP specifies that interaction with the caller is needed to complete the call. Specifically, it requires an announcement to be returned to the caller saying 'If you want investment services please enter 2. For insurance services please enter 3'. And it requires a digit to be received by the Intelligent Peripheral.

The SCP therefore sends a CCSS7 signalling message to the SSP asking it to connect the Intelligent Peripheral to the caller and start the announcement, as shown. On receipt of the digit entered by the caller, say '2', the Intelligent Peripheral sends this supplementary information to the SCP via the SSP which also releases the IP from the caller. The SCP responds with a signalling message advising that the destination number for this call is 01234 654cdf, the number of the nearest high street branch's investment services team. At this point the SSP resumes basic call processing using 01234 654cdf as the destination number and the call is completed in the usual way.

Note that in general transit exchanges do not have SSP functionality (though they may). The power and flexibility of CCSS7 allow functionality to be located on a cost/performance basis.

Centrex, an example of a VPN

Centrex, which has been widely used in the USA for a number of years, is perhaps the best-known example of an Intelligent Network service and

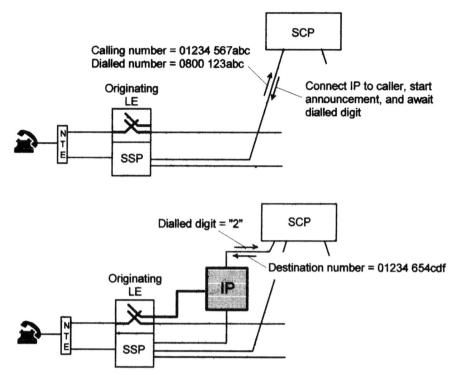

Figure 6.12 An IN service with user interaction

provides a flexible alternative to the traditional private network—in effect a VPN. Whilst traditionally a company's private voice network consists of a number of PABXs interconnected by leased lines, Centrex does away with the PABXs. Instead, the functionality of the PABXs is provided by the service provider's IN.

At each customer site access to the public network—the IN—is provided by means of one or more multiplexers which concentrate the site's telephones onto IN access circuits as shown in Figure 6.13. These multiplexers act as concentrators in that there are fewer voice circuits between the multiplexer and the IN than there are telephones on the site. It would be a very rare event for every telephone to be in use at the same time, and the size of the access circuit group is dimensioned using well-known teletraffic principles.

Centrex offers the corporate customer a similar service to the traditional private network of PABXs, including customised numbering schemes (typically of seven digits), but tends to be more flexible. For example, the numbering scheme can include small remote offices having only a few telephones, or perhaps only one as shown in Figure 6.13. Telephones in such remote offices would have both a seven-digit number (say) and a PSTN/ISDN number. The jargon for this is that telephones can have both on-net and off-net numbers.

In addition Centrex offers uniformity of features in supplementary services (such as call diversion, ring-back-when-free, etc.) across all sites, including

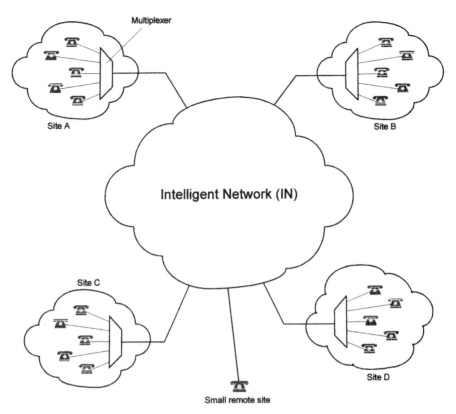

Figure 6.13 Centrex

small remote offices. With networks of PABXs such uniformity is rare if different makes of PABXs are involved as they usually are. Whilst historically the most advanced features have usually been available in PABX implementations before they have been available in Centrex offerings, this is now changing as IN implementations mature.

'Green field' situations where a company could choose between a private network of PABXs and Centrex are very rare, if they exist at all. Every company has a history and therefore a legacy, usually a PABX network. This cannot be simply discarded, and moving its operations at once from a PABX network to Centrex would not in general be acceptable. It would put too many eggs in an untried basket. So in practice Centrex tends to be used in combination with PABXs in what is known as a 'hybrid' network as shown in Figure 6.14.

This requires additional functionality in the IN to integrate thePABX functionality with the Centrex service. Any Centrex service that does not support hybrid working will not sell very well. A possible drawback of a hybrid network is that there may be some loss of transparency in the supplementary services as implemented by the PABXs and by the IN. Over time, as PABXs reach the end of their economic life, if the Centrex service proves to be reliable it is likely that a company will migrate towards a wholly

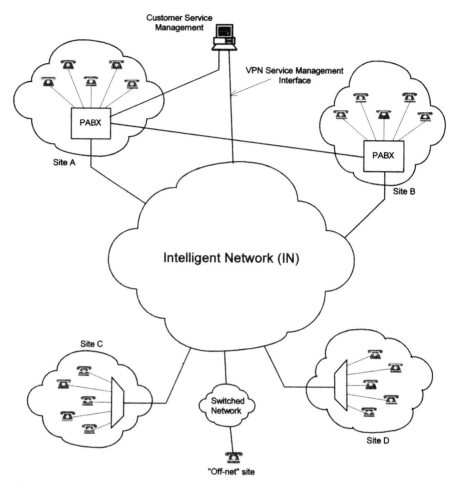

Figure 6.14 A hybrid network

Centrex solution. Experience in the USA is that most large companies use more than one Centrex supplier in order to avoid becoming locked into one supplier and to keep the Centrex service providers 'on their toes'.

There are basic economic differences between private networks and Centrex. Private networks have high capital costs for the infrastructure but usage costs are fixed and known in advance. With Centrex, however, the capital costs are small but running costs include a usage element that may be difficult to forecast. However, as we have seen, Centrex offers considerable agility to track changes in a company's operating and trading environment, whilst the high capital investment bound up in a traditional private network tends to militate against this. It should also be remembered that large companies tend to be multinationals, and the scope for cost savings on international leased lines can be considerable. The main VPN suppliers tend to have international coverage, often based on a consortium of service providers.

At this point we will change terminology from Centrex to VPN since the features we are going to discuss tend to go beyond those traditionally associated with Centrex. But the distinction is not important and in practice is likely to reflect marketing considerations as much as technical ones.

Access to VPN services may be obtained indirectly via one or more switched networks as shown in Figure 6.14. The terminology is that sites directly connected to the IN are 'on-net', whilst those with indirect access via another network are 'off-net'. Indirect access may be implemented in a number of ways. An off-net user may dial a pre-allocated access code, typicaly of four digits. This would route the caller to an access port on the IN, which would then implement a dialogue with the caller to obtain an authorisation code. After receipt of the authorisation code the caller would be treated as a regular part of the VPN (for that particular call: the authorisation code is used on a per call basis). 0800 Freefone access can be regarded as a particular case of a pre-allocated access code. Indirect access can be used to include public payphones and mobile telephones as part of a VPN.

Clearly telephones that gain indirect access using an authorisation code may also be used independently of the VPN (simply by omitting the access code). Another option, however, is to use 'hot-line' access whereby a remote telephone is dedicated wholly to the VPN. When the handset is lifted on such a telephone it is immediately connected through the access network to an access port of the VPN. In this case an authorisation code is not needed since the Calling Line Identity of the caller is passed to the VPN by the access network (in effect there is 'point of entry policing') and the telephone is usually regarded as on-net.

The last point illustrated in Figure 6.14 is that of customer management of his VPN. In the traditional private network of PABXs the company has complete control of his network. This may be seen as an undesired burden or as an important business control. But in any case it creates an expectation on the part of the company that it should have a degree of direct control over 'its' network, and with the increasing competition between VPN service providers, the degree and ease of control offered to the VPN customer is an important differentiator. Typically a VPN customer will be offered remote control of a number of service aspects, including:

- the numbering plan;
- authorisation codes and passwords;
- call routing (for example, where a call is routed partly in the VPN and partly in another network it may be important to control the VPN routing to minimise call costs in the other network);
- call screening (e.g. barring of international calls or premium rate services, perhaps with override authorisation codes).

In addition the VPN customer would typically be provided with on-line access to reports, including:

- network costs—the customer wants no surprises in budgeting;
- network performance—to check that service level agreements are being met by the service provider;
- details of usage, such as calls made out of normal business hours;
- network traffic reports, typically produced daily, weekly, monthly, on-demand, or whenever preset thresholds have been exceeded.

General

Summarising all this, the basic ideas of the Intelligent Network are:

- to separate basic call control from customised aspects of call control;
- to do basic call processing in the Service Switching Point (SSP);
- to do the customised aspects of call control centrally in the Service Control Point (SCP);
- the detailed information relating to a corporate customer's service is held in the SCP (usually);
- the Service Switching Point is a modification of the existing exchanges;
- the Intelligent Peripheral (IP) provides specialised resources.

The whole point of this centralisation of intelligence is to ease the otherwise intractable problem of creating and managing the customised services for the corporate customer. In this way we can realise the full potential of stored programme control of switched networks to provide services tailored to the requirements of individual customers quickly, flexibly and reliably. To achieve this in practice additionally requires effective means for creating and managing these individually tailored services.

6.5 SERVICE MANAGEMENT AND CREATION IN THE IN

Service management

Effective service management requires a good Service Management System (SMS). This is used for:

- adding new customers and services;
- synchronising changes to service data and Service Logic Programmes across the SCPs (as we have seen, there will always be more than one, possibly quite a few);

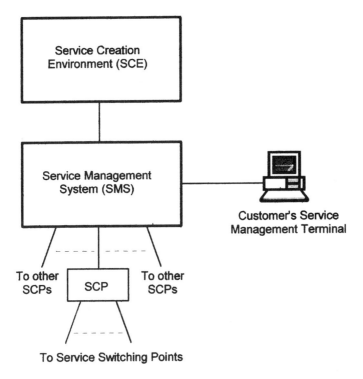

Figure 6.15 Service creation and service management in the IN

- administering the database(s);
- reloading service data and software following an SCP failure and managing its return to service;
- bringing new SCPs into service;
- network and service surveillance.

This list is certainly not exhaustive, and for the service provider to remain competitive the Service Management System itself will be the subject of timely upgrading and enhancement to reflect new demands, threats and opportunities.

The place of the Service Management System in the scheme of things is shown in Figure 6.15.

Historically companies will generally have invested heavily in their own private network infrastructure. Their control and management of this network infrastructure has traditionally been both comprehensive and direct. If they are to be persuaded to forsake their private network for a VPN approach they need to feel that they are still in control. An important element of the SMS therefore is the management interface and capability it gives to the customers. They want, and are used to having:

- up-to-date information on performance;
- the ability to change telephone numbers within 'their' network;
- the ability to change access authorities, such as barring international calls or calls to premium rate services.

But customers' on-line access to the Service Management System has to be very carefully controlled to ensure that they cannot, deliberately or unwittingly, affect somebody else's VPN (or indeed, the service provided to the 'public'). Firewalls are therefore used to prevent customers from getting access to any capabilities they are not entitled to (or have not subscribed to).

Service creation

We have already suggested that the speed at which a service provider can produce a customised solution to corporate customers' needs is an important aspect of his competitiveness. An effective Service Creation Environment (SCE)—basicaly the set of tools used to create and test new or customised services—is therefore a key requirement for success. And once you have made the 'sale' you cannot afford to lose it because of poor in-service performance. there is also considerable scope for a flawed IN service to wreak havoc in the public services provided on the same network platform. So service creation needs to be not only fast, but also robust and accurate.

The Service Creation Environment is likely to use object-oriented methods, a powerful graphical user interface, and an Application Programming Interface (API) that reflects the Service Logic Execution Environment. Ideally, it will support the complete service lifetime, embracing requirements capture, service specification, service demonstration, design and development, service trials, software release control and deployment, and in due course service termination. Figure 6.16 shows a typical service creation process.

6.6 CENTRALISED vs DISTRIBUTED INTELLIGENCE

Our treatment of the IN has so far been based on centralised intelligence, since in practice it is virtually impossible to control a distributed intelligent database spread over hundreds, or even thousands, of locations, at least when the service information is changing frequently. But ITU-T in their wisdom have created IN standards that also embrace distributed intelligence. One of the reasons for this is that, despite the problem of managing such distributed intelligence, there are situations in which centralised intelligence just does not make sense. an obvious example is SMDS, discussed in some detail in Chapter 4, which provides a high-speed, wide-area connectionless switched data service.

Being a connectionless service, every SMDS packet is treated in isolation—

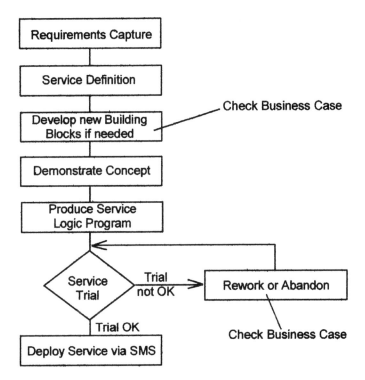

Figure 6.16 The service creation process

there is no sense of a call being set up or of sequence in respect of the packets. In effect every SMDS packet can be regarded as a 'mini-call'. So if we add centralised intelligence to an SMDS network, as shown in Figure 6.17, we would have to make reference to it on a packet-by-packet basis. The problem is that it would take too long. The cross-switch delay incurred by an SMDS packet would typically be less than a millisecond (remember that since the control information in an SMDS packet is concentrated at the beginning of the packet, a switch can begin to send a packet on an outgoing link even before the whole packet has been received). But it would take several tens of milliseconds for a routing enquiry to be referred up to the central database and for a response to come back, which is clearly untenable.

We have seen in Chapter 4 that Closed User Groups, an important feature enabling VPNs to be built on a public network, are implemented in SMDS using address screening tables. To create an SMDS VPN for a company therefore requires the network operator to manage the entries in the address screening tables in all SMDS switches that directly serve that company. In practice this would be done by maintaining a central 'map' of the VPN and downloading incremental changes to the address screening tables in the SMDS switches to reflect changes in service requirements. This would reap many of the benefits of centralised intelligence whilst keeping the fast response time of localised intelligence.

The saving grace is that SMDS networks are still comparatively small and it

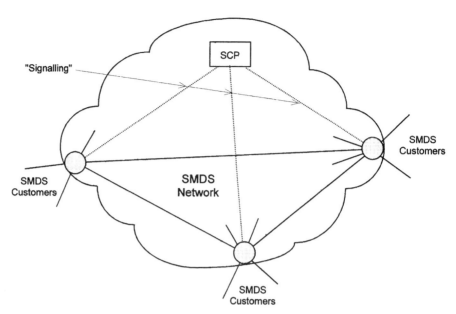

Figure 6.17   Centralised intelligence with SMDS just does not work

is still a practical proposition to manage the corresponding distributed database. If it is true, as we argue, that telecommunications will in future involve more and more customisation of the services offered to the corporate customer, and that only VPNs offer this with the flexibility needed for companies to respond quickly to new situations, then it follows that this inability of connectionless services such as SMDS to scale is a constraint its future growth. Alternatively, of course, this constraint will not arise if the science of managing large distributed databases keeps pace with the growth in connectionless wide area networks.

As we have already seen, ITU-T distinguishes between a function and its location. Figure 6.18 summarises the main functional elements used to build Intelligent Networks and where they may be located.

Though we have not considered it in the above description, there is also a Call Control Agent Function (CCAF) which defines a subset of the Call Control Function (CCF) that may be implemented remotely from the CCF—typically a CCAF would be located at a Local Exchange with the full CCF implemented at the nearest Transit Exchange. And there is a Service Management Agent Function (SMAF) which defines parts of the Service Mnagement Function (SMF) that may be implemented remotely from the SMF.

The mandatory entries in Figure 6.18 reflect the centralised view of intelligence as we have developed it for the IN. The optional entries embrace distributed intelligence.

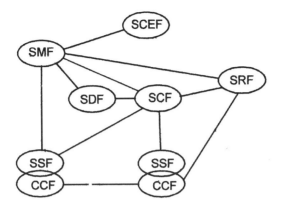

CCF Call Control Function
SCEF Service Creation Environment Function
SCF Service Control Function
SDF Service Data Function
SMF Service Management Function
SRF Specialised Resource Function (The IP)
SSF Service Switching Function

	SSF/CCF	SCF	SDF	SRF
SSP	Mandatory	Optional	Optional	Optional
SCP		Mandatory	Optional	
SPD			Mandatory	
IP	Optional			Mandatory

Figure 6.18 IN functional elements and their location

6.7 SUMMARY

This chapter has built on the idea of separating the network into a switched information subnet and a signalling subnet, and introduced the idea of adding intelligence to the signalling subnet in order to fulfil the original promise of stored programme control. In many ways this can be thought of as putting the operator back into the network. The most intelligent network technology every deployed was the human operator sitting at a manual switchboard. In the very early days of telephones it was possible for the small group of operators looking after a small town to know everyone in the town who had a telephone and their business. If someone called the doctor, the operator would typically know that he was delivering Mrs Smith's baby at No. 16 London Road, and could redirect the call accordingly.

The separation of the switched information subnet and the signalling subnet was a feature of the manually-switched telephone networks of the nineteenth century. Indeed, the separation was not only logical but physical. The switched information subnet consisted of the manual boards (panels of arrays of jack sockets into which the operator inserted the appropriate plug-terminated cords to effect the desired connection), whilst the signalling subnet consistent of the operators (who sat at adjacent exchanges and communicated with each other by means of 'order wires', equivalent to the CCSS7 Message Transfer Part). The customer's signalling interface with the network was the most natural imaginable, i.e. free-format speech.

It is interesting to note that the intelligence and flexibility of this control mechanism could be easily subverted and that this was what led to the development of automatic telephony. The story, perhaps apocryphal, is that Almon B. Strowger, an undertaker by trade and the inventor in 1898 of the first viable automatic switches for telephone, was motivated to design his switch because he found out that a rival undertaker had bribed an operator to divert calls, and thereby business, from Strowger to himself! Perhaps what this illustrates is that, whilst we may get the odd malfunction in the user-plane, it is in the control-plane that the real power lies to create havoc, and to develop the power of the C-plane to fulfil the full potential of stored programme control requires a well-defined architecture if the havoc is to be avoided.

This chapter has:

- developed the notion of separating the signalling and switched information subnets;

- explained how intelligence is added to the IDN/ISDN to form the IN;

- introduced the main ideas and terminology of the Intelligent Network;

- shown how the IN enables customised network services to be constructed;

- developed the idea of 'agility' as a key business requirement; Virtual Private Networks (VPNs) provide equivalent agility in terms of corporate telecommunications services.

The idea arises from this of service providers offering IN services without having any kit of their own, but using the network infrastructure of other service providers—in effect an advanced form of resale. There seems little doubt that such service providers will arise, probably sooner than we think. History has shown that if customers want something and technology can provide it, then sooner rather than later the necessary regulatory framework will be developed to facilitate it.

But the approach described in this chapter for implementing customised networks is not the only one. It is the approach spawned by the telecommunications culture designed to provide the flexibility and scalability needed by many corporate customers. The next chapter outlines a corresponding approach, the intranet approach, developed by the IT culture.

REFERENCES

General

BT Technology Journal, **13**, No. 2, April 1995—Special issue on network intelligence.

Knowles, T. *et al.* (1987) *Standards for Open Systems Interconnection.* BSP.

Spragins, J. *et al.* (1991) *Telecommunication Networks, Protocols and Design.* Addison-Wesley, London.

Standards (note that the following lists are not exhaustive)

ITU-TS
CCITT Signalling System No. 7

Q.699	Interworking between Digital Subscriber Signalling System No. 1 and Signalling System No. 7
Q.700	Introduction to Signalling System No. 7
Q.701–Q.710	Message Transfer Part (MTP)
Q.711–Q.716	Signalling Connection Control Part (SCCP)
Q.721–Q.725	Telephony User Part (TUP)
Q.730–Q.737	ISDN supplementary services
Q.750–Q.754	Signalling System No. 7 Management
Q.761–Q.767	ISDN User Part (ISUP)
Q.771–Q.775	Transaction Capabilities
Q.1200–Q.1290	Intelligent Network
Q.1400	Architecture Framework for the development of signalling and OA&M protocols using OSI concepts

7

A Computing View of the Total Area Network

One of the hottest frontiers in technology is for software that will enable people to co-operate across national and economic frontiers. If you are not on the net, you are not in the know.

Fortune Magazine

Despite having no first-hand experience of the act, the authors have been assured by many friends and colleagues that there is more than one way of skinning a cat. In this chapter, we take a different view of a Total Area Network and illustrate how a similar end can be met from a completely different angle.

In particular, we explain the various steps in putting together an Intranet, that is a Virtual Private (Data) Network based on Internet technology. The main aim in doing this is not so much to illustrate the components and applications that comprise an Intranet. It is more to show the approach taken and what the main concerns are when a computer-centric view is taken of a Total Area Network. The essential point is to show that the telecommunications approach builds on centralised intelligence, but the net approach, in contrast, distributes intelligence. This distribution is partly in the servers and partly in the applications—which are analogous to service logic aspects explained in the previous chapter. The overall effect is that, instead of adding computers to communications, here we add communications to computers.

We start by explaining some of the basic elements of an Intranet and the technology that is used to assemble one. Following this, we look at some of the applications that would typically be hosted on the network and how they would be used. To close, some of the key issues with this type of Total Area Network are discussed along with a few of the practical implementation details.

7.1   INTRANET BASICS

The emphasis of this book in earlier chapters has been very much on network technology. Given that the thing that concerns most people is not the network technology *per se* but what it is actually used for, we now start to look at how applications fit onto a network. In particular, we explain what sort of computing issues come to the fore as services are overlaid onto a Total Area Network to create an Intranet.

Perhaps the most concise way to define an Intranet would be to say that it is the deployment of Internet technology to meet the needs of particular group or organisation. We should, at this point, make one thing very clear—even though it is built on the same underlying technologies as the Internet, an Intranet is in fact quite different. The fact that it is built to the requirements of a prescribed customer base makes it a technology-specific example of a virtual private network. This means that performance and quality of service guarantees need to be designed in so that the customer feels that they are getting value for money. The basic set-up of an Intranet is shown in Figure 7.1.

In reality, there would be numerous computers, servers and local area networks connected together. Typically, an Intranet constructed along the lines of Figure 7.1 would provide a range of information services to user terminals. News feeds, mail, file access and on-line references would be provided along with a host of other information services (Frost and Norris 1997). In many companies, the Intranet provides the main source of working material, the vehicle for co-operative projects and the preferred reporting route.

Because of the well-established components used for an Intranet, the way in which it is put together is not in line with the picture already built up for a telecommunications oriented Total Area Network. Nonetheless, there are many similarities and we begin in this section to build an Intranet from the ground up.

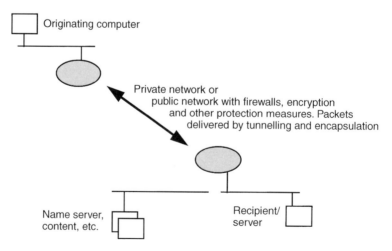

Figure 7.1   An overview of an Intranet

First, the main components. The basic set that is needed to put together an Intranet is:

- user devices capable of communicating with the network and displaying the information served from it—typicaly a PC but could also be a Network Computer or NC;

- a browser, such as Netscape Navigator or Microsoft Internet Explorer. This provides a familiar and standard user interface (or client);

- computers (or servers) to provide the information and services. These are usually protected by firewalls to ensure privacy;

- a set of established protocols that link the clients to the servers and allow information and service to be delivered to the standard client.

Behind the technology used lies a basic philosophy (common both to the Internet and to Intranets)—that the available components should be deployed in the well-established client/server configuration, one that has proved effective for many distributed computing applications.[1]

That is where the commonality between the Internet and Intranets ends. Because it serves the needs of a specific organisation, an effective Intranet must provide levels of performance, security and manageability that suit the needs of that organisation and that cannot be guaranteed over the public Internet.

By way of illustration, most people's experience of the World Wide Web is of 'recreational surfing' in a setting that is often indistinguishable from anarchy. Organisational users who rely on an Intranet to do their jobs need a more structured environment and thus have an entirely different set of requirements.

Thus, any Intranet solution must be designed to recognise the existing organisational infrastructure and allow people to interact and work the way they normally would, while also providing opportunities to enhance and redefine how things are done. Crucial aspects for any corporate or community Intranet are:

- the privacy and integrity of the information stored on and transferred across the Intranet;

- the structuring of operational data, which has to be customised to suit the organisation using the Intranet;

- the predictable nature of the service, in terms of the speed with which both services and information are delivered;

- the ease with which the Intranet can be reconfigured to accommodate changes, both in users and in how services are managed.

The extent to which these key criteria are met is determined by the way in

---

[1] Intranets are, as *PC Week* recently defined them, 'client/server done right and done fast'.

which an established set of components are deployed. In other words, the suitability of an Intranet depends on the quality of design. And, as should be clear from our story so far, this means the marriage of traditional telecommunications concerns such as capacity, latency and throughput with the time-honoured computing issues of presentation, formatting and data modelling. A good design will combine these and also attend to the overall manageability and security of the Intranet.

Before continuing, a couple of myths to dispel. First is that an Intranet requires a connection to the global, public Internet. It can exist as a separate entity, even if many Intranets do take advantage of the Internet. The second myth is Intranets are inherently insecure. There may be a great range of vendors of Internet technology but the way in which any Intranet is put together governs its security—the level is a matter of design, not an inherent property.

7.2   APPLICATIONS AND OPERATIONS

A real Intranet is a complex entity that has a significant number of separate co-operating elements. Looking back at Figure 7.1, we can see how each component fits into the overall network, and hence service.

To start with, at the local or user end, there are usually lots of PCs and workstations connected to Local Area Networks. As already mentioned, each user screen is equipped with browser software. This provides a standard 'window' on the Intranet and is the normal access route to all services.

Moving in from the local environment, there is the link from the Local Area Network to the transmission core that carries messages, files and other information across the enterprise. Typically, this would be provided by a router connected to an SMDS or ATM point of presence.

Lastly, there is the core itself—the Wide Area Network. This provides the transport mechanism between all of the connected elements. This part is usually shown as a cloud as it is often capacity bought in from a third party—it may, in practice, be a set of private wires or capacity on a commercial data service such as SMDS or ATM (e.g. using the BT Cellstream service). In some cases, the core may even be provided using the public Internet service.

The user communicates with the server using standard LAN mechanisms. Within the packets that constitute local communications are embedded the Internet Protocol or IP packets (which themselves embed application information carried across the Intranet). In their turn, the whole lot may be carried inside the packets used in the ATM or SMDS networks used for wide area connectivity.

At the point of interconnection to any public network there is usually some form of firewall to police the packets passed to and (more importantly) received from the outside world. All of the Intranet services—naming and authentication, etc., as well as the information that is on offer—are provided from servers connected at local ends.

In any Intranet, there are a number of links that have to be made before the whole thing works. The various steps in communication are best explained through an example.

Let's assume that we have our end user's computer, connected to a Ethernet LAN. The physical address associated with the computer, in this case, is a 48 bit or 6 byte identifier known as the Ethernet's MAC address. When the user wants to communicate, their computer broadcasts a packet (known as an Address Resolution Protocol or ARP packet) within an Ethernet frame. As this is an Intranet and IP is used as the networking protocol, the packet contains the destination IP address of the intended host.

The hardware location of the addressee is not known to start with, so the hardware address field in the broadcast message is set to zero. The packet is broadcast to all devices on the Ethernet LAN. Once the broadcast packet is received, the addressed device can fill in its hardware (MAC) address. It then transmits an ARP reply in which its physical address is inserted in the field previously set to zero.

Hence the logical IP address of the intended recipient is associated with its physical MAC address. Once this step is complete, packets can be sent between the two without delay; a logical path now exists.

Now that the end points of the communication are established, something else is needed to allow the processes at either end to interact (and for the service to be provided). This is where the Transmission Control Protocol, TCP, fits. The TCP header contains source and destination port fields that identify specific applications that reside on the server. For instance, port 80 is the 'well-known number' reserved for hypertext (HTTP based) applications, port 25 for Mail (SMTP), port 23 for remote login (Telnet), ports 20 and 21 for file transfer (FTP). By the end of this step, the client and server know how to address each other and what service is being delivered.

The above description reflects the 'layered' approach that is often taken in building communication services. Notionally, each of the sequential steps we have run through supports the next one and there is a stack of protocols that co-operate in providing the complete service. This is illustrated in Figure 7.2.

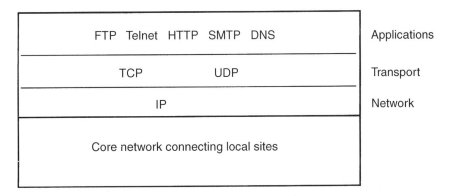

Figure 7.2 The protocol stack used in an Intranet

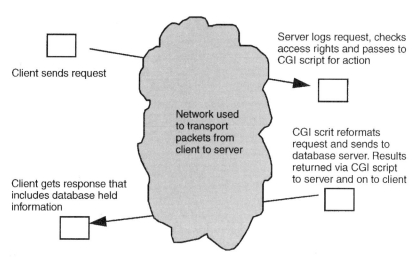

Figure 7.3 A complex transaction over the Intranet

At the base we show the network connection that allows subsequent protocol interactions. The next layer sets up basic transport between the computers that are connected to the network. Finally, the specific applications that the user perceives as network-provided services are enabled.

Together, the ARP, IP and TCP protocols give us the wherewithal to associate devices with each other and then ensure that they communicate appropriately and deliver a specified service. A little more about these, and other key Internet protocols, in the next section.

Once the supporting cast have done their job, the browser software loaded onto the client computer can initiate a session with a server, typically using the HTTP protocol. The ensuing client/server interaction would usually be a straightforward request for a file or page of information (West and Norris 1997).

Alternatively, a more complex service, such as a database access (e.g. to look up a phone number) may be provided. In this instance, the client request is passed by the server to another application via the Common Gateway Interface or CGI. This is illustrated in Figure 7.3.

In practice, CGI provides a script that reformats information from any system that the Intranet server can see, so, for instance, databases can be queried and the requested information presented to the user through a browser as if it were a sourced from the server itself. There are several other ways in which advanced services can be provided over an Intranet. For instance, mobile code (that is software in a language such as Java that can be downloaded from a server to a client and executed locally), delivered via the Intranet, can be used for many purposes.

Hence, end users have access to a wide range of information through a common interface. The fact that a network sits between the user and that information is, to all intents and purposes, invisible (or at least should be). This 'transparency of location' is one of the ideals of any distributed system (Norris and Winton 1996). It is effected by a combination of the common access techniques (browser interface, associated protocols) and wide area network that transports data between client and server as if they were co-located. CGI suits lower volumes but when company-wide scaling is required, a more robust option is needed. For instance, a staff directory, which may take millions of hits per month, might store records on an Oracle databases and use their application servers for access. In this instance, the interface to the server is engineered to cope with a high volume of transactions, rather than the simple protocol and script interface shown above. Distributed system technology (in this instance, CORBA) allows many servers to support an application and hence provide the scalability to cope with sophisticated company-wide utilities.

7.3 THE TECHNOLOGY USED

Between the network that takes the distance out of information and the information itself, our Intranet relies on a number of established protocols, as already mentioned. These are familiar to most Internet users but, for completeness, here is a brief description of the main ones:

- IP—the Internet Protocol—is a connectionless packet protocol with a structured 32-bit address field. Each IP packet carries source and destination address plus length, version check and other safety information (e.g. the packet's time to live, if misdirected). A complete IP packet can be up to 65 kbytes in length. With LAN datagrams of the order of 4.5 kbytes (Token Ring) and 1.5 kbytes (Ethernet), fragmentation of the IP packet can occur but this is transparent to higher level protocols. The latest definition of IP—IPv6 or IPNG—has extended capability but is not yet in widespread use.

- TCP—the Transmission Control Protocol—is a connection-oriented protocol which sits above IP to provide end-to-end connection control. Like IP, it has a packet structure and this allows a number of application processes to be mixed within a single data stream. The concept of well-known ports (80 for HTTP, 23 for Telnet, etc.) is used to associate data in the packet payload with the application that it is intended for.

- DNS—the Domain Name Service—provides a hierarchical mechanism for converting mnemonic addresses (fred@ibm.com—also known as 'dot' addresses) into 32-bit IP addresses. So 142.32.156.19 (which is used by the network) is resolved into mnorris@iee.org, which is the user-recognised address.

- HTTP—the HyperText Transfer Protocol—provides the mechanism for getting hypertext information from a server to a browser. Closely linked with HTTP is the Uniform Resource Locator or URL, which identifies the server to be used. The most familiar URL is http://www.server.com (which is no more than a normal Internet dot address prefixed with a qualifier to say that HTTP should be used to handle the transaction).

- HTML—the HyperText Mark-up Language—is the basic authoring language used on the World Wide Web. Along with the many tools to support it, HTML allows pages of hyperlinked multimedia information (i.e. linked text, images and video) to be created and managed.

- Web browsers—such as Netscape Navigator and Internet Explorer—allow access to the HTML coded hyperlinked information. In addition, they provide integrated E-Mail facilities which provides the foundation for workflow and groupware solutions.

- NNTP—Network News Transfer Protocol—provides the basis for threaded discussion groups like those found on many of the commercial on-line services (e.g. Compuserve).

- IRC—Internet Relay Chat—is an open discussion group facility that provides the basis for real-time, text-based collaboration and whiteboards as the basis for real-time, graphical collaboration.

The list could go on to incorporate a raft of others, from those used for special purposes (such as SNMP for management) to emerging technologies such as ActiveX and Java that provide portable applications that can be transported across a network.

Most of those listed above come as standard offerings on a wide range of computing equipment. The technical detail of all of these protocols can readily be found in the Internet technical definitions, known as RFCs (or Requests for Comment), available through the Internet Society Web pages (www.isoc.com).

7.4 SECURITY

An Intranet serves a specific closed user group and it is increasingly the case that valuable information is hosted on the net. Hence security is a key issue (as it is in any shared medium network) and needs to feature high on the agenda for both the designer and user of an Intranet.

There are a number of security risks to the user of an Intranet. The main ones are:

- eavesdropping—the obvious problem on any shared medium network;

- connectivity—there are many, many places from which an attack on a system may originate;

- forgery—with little effort needed to produce a stream of fake bits and bytes (which could represent an electronic funds transfer between banks), the temptation to indulge in forgery is obvious;

- unforeseen use—where pieces of software get used in ways that their original designers never envisaged;

- complexity—in any complex system it is easy to overlook things which might compromise the overall security.

With so many of the components from which the Intranet is constructed being standard, it is the way that they are organised and managed that has to be attended to. In particular, there are five vital aspects that need to be catered for in a security design:

- integrity—ensuring that the system remains in a consistent state;
- privacy—the ability to keep secrets;
- identification/authentication—the ability to determine who (or what) is trying to make use of the system;
- authorisation—the process of deciding exactly what actions the user is or is not allowed to perform (i.e. what access rights and privileges they should have);
- auditing—to provide accountability for and traceability of actions so that problems can be traced to source.

It is important to stress that, in achieving these, security is not an absolute concept but a matter of degree. It is impossible for any system to be totally immune to attack (or misuse). The amount and level of security in a design should be proportional to the value of the information or services provided by that system. Within this it is useful to differentiate between network security, data integrity and application security. Each poses a different set of threats to be addressed. For instance, application security is usually tackled with user passwords and secure access cards.

One of the most effective and widely used strategies for preserving

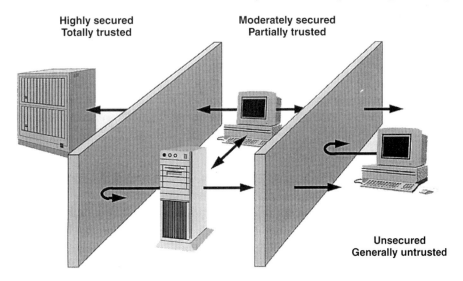

Figure 7.4 The concept of a firewall

network security is to use a 'firewall' system (Cheswick and Bellovin 1994). Just as a real firewall is specially built barrier which stops the spread of a fire within a building so a computer firewall tries to limit the extent of damage to computer security. The basic idea is that machines and networks on the 'inside' of a firewall are trusted, those 'outside' are generally untrusted (Figure 7.4).

All communication between machines inside and outside goes through the firewall (which, in practice, is usually a special purpose computer such as the commercially available PICS from Cisco or Netra from Sun). It is the job of the firewall to monitor and filter all traffic passing through it and only to allow through known communication corresponding to certain well-defined services or to trusted external systems.

Machines on the inside of a firewall have a reasonable degree of trust of other machines within the firewalled network and so have to apply fewer controls themselves. Sections of the internal network may themselves employ firewalls to restrict the domain of trusted machines even further.

Firewalls usually concentrate at the level of network transport (i.e. they look at the content of the packets used to transfer data between computers). However, the same idea is also applied to higher-level services with good effect. A service may be shielded by a proxy server, where the proxy presents the same interface to the outside world as the real service but instead of performing the requested operations itself it filters out undesirable requests and only passes approved requests on to the real server.

A point made at the start of this section was that there is a lot more to security than filtering out untrusted packets. The key point to close on is that a known level of security can be designed into an Intranet and this is one of the points of differentiation between a good and a bad implementation.

7.5 MANAGEMENT

As network service and on-line information systems become increasingly central to people's working environments, so their failure becomes less acceptable. For many years, the operational support systems that surround large networks such as the Public Switched Telephone Network have attracted considerable investment of both time and energy. A similar level of investment needs to be put into keeping an Intranet fit for purpose.

There are a host of things that need to be done to keep any network operating as it should. A few of the main issues that most people recognise need to be coped with are:

- getting an 'end-to-end' view of the Intranet. This is not always easy when the component parts have been built to be managed as isolated units (if at all);

- configuration control—because it is very difficult to stop machines and software being introduced into any large network in fairly haphazard manner;

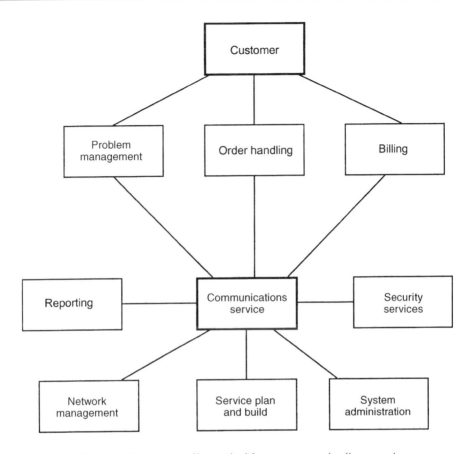

Figure 7.5 The 'service surround' needed for a communications service

- information overload—because a single error may affect a number of components each of which may adopt its own error handling and reporting strategy.

This list can readily be extended. The key point, though, is that a suitable 'service surround' needs to be put over the Intranet so that users can be added and removed, services configured, charges raise and faults fixed. As implied above, the concept of 'service surround' is very much 'business as usual' for telecommunications service providers (West, Norris and Stockman 1997) and this is reflected in Figure 7.5.

Just how all of these capabilities are provided is, again, a matter of design. For instance, a typical Intranet might well deal with the network management aspect by setting up a regime that first checks that the required routes are there (e.g. by pinging and polling routers) and subsequently (by accessing status information on servers) that performance is satisfactory (i.e. not too many packets are being dropped, etc.). The next step would be to gather faults and alarms and correlate them so that root causes are revealed. This may involve trading trouble tickets with third parties and would invariably call for the installation of a specialist fault management system.

We have already explained a little of the detail behind Intranet security. Each of the service surround areas pictured above has its own established standards and practices. For the rest of this section we will focus on one of these—network management.

The basic facilities needed to manage any network are the abilities to monitor and control the operation of remote components such as servers, routers and communications links. Obviously to perform any sort of remote management physical intervention is impossible. This is where the concept of management agents comes into play.

An agent is effectively a special kind of server which provides a view to any interest clients, or managers, of a particular resource. A resource is basically anything which can be managed. It might be a computer or a printer or even a local area network. It can even be purely a software artefact, such as a print queue.

Agents provide a standardised view of resources to the outside world. They usually do this in terms of a set of one or more managed objects which encapsulate the resource. It is a purely internal issue as to how the agent actually interacts with the real resource. A set of such objects for a resource or group of resources is described as a Management Information Base or MIB. A MIB is effectively the collection of interface definitions for the managed objects.

Each managed object has a series of attributes which describe its characteristics. So a print queue might have an attribute which describes the number of currently queued print jobs or whether the queue is currently enabled or disabled. Remote operations (such as 'get', 'set', etc.) can be performed on objects and their attributes. Commonly a *get* operation reads the value of an attribute. Some, but not all, attributes may be remotely *set*—which usually causes some resulting change in the behaviour of the resource. So to carry forward the print queue example, changing the status from 'enabled' to 'disabled' should cause the queue to stop accepting requests.

The real challenge of network management, though, is not interacting with individual elements. It is about getting a view of the overall state of a system at a particular moment. The easiest way of achieving this and the one most commonly used is polling.

This is the process of repeatedly asking agents for their current state. So a management system which polls its agents every few minutes should have a picture of the system which is reasonably accurate—provided that the rate of change within the system is not too high. Changes between the current and previous states can be detected and highlighted as necessary. It should be pointed out at this point that a balance has to be struck between gathering management information and efficient operation—it does not take long for management traffic to eat into network capacity.

So, although effective, polling is not always particularly efficient in practice. The usual alternative is the use of events and notifications. When an agent finds that the state of a mnaged resource has changed it sends out an unsolicited message called a notification to interested parties. The notification contains information about the event which occurred. So a printer running out of paper might cause a notification of an 'out-of-paper' event to be generated and sent out.

Event notification is a much more attractive model in many circumstances. No resources are wasted in continually asking for status reports when nothing has changed. Information is also likely to be more timely and fresh because events are processed as they happen rather than at the next poll cycle. On the other hand, without any control over event generation it can be all too easy for a manager to get flooded with a rush of notifications, particularly if a catastrophic failure—such as a core network link being lost—causes a large number of agents to generate events simultaneously.

For this eventuality, the solution usually adopted is filtering. In essence, this means that only events of a selected type are sent to the manager with others being discarded. Associated with filtering is correlation, whereby events related to the same root cause are marshalled into one event.

Overall, the event-based management view is very powerful, particularly when combined with other techniques such as correlation and escalation. For example, an occasional hiatus in network performance may be insignificant, but a fault that occurs repeatedly could be much more serious. Administrators would like not to be bothered by the former case while being confident that they will be alerted of the latter. The main disadvantage of event-based management is that it usually requires somewhat more sophisticated and complex manager and agent implementations.

As one would expect, there are some standard protocols that can be used for network management, the best known being the Simple Network Management Protocol (SNMP). As with the other protocols described earlier, SNMP and its like are Internet standards.

Also, there are quite a few commercially available tools that help with these and other network management tasks. The challenge before the Intranet designer is to ensure that the tools chosen can cope with volumes, don't need too much integration before they work together and are relevant to the needs of the users.

Once you have your network management under control, there is billing (either for real cash or for internal accounts), provisioning, capacity planning, etc. to be taken care of. The key message of this section is that the service surround to an Intranet is at least as important and as complex as the Intranet itself.

7.6 PRACTICAL OPTIONS

With increasing use of images, sound and video within information systems, it is important to ensure that enough bandwidth is provided to support the amount of information that is likely to be transferred. This means that the available capacity between clients and server (e.g. the committed information rate in Frame Relay terms) should be planned to cope with expected demand (and then doubled!).

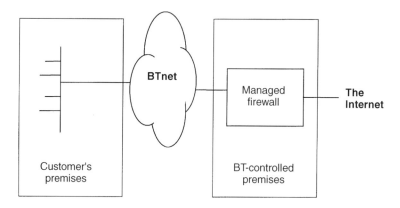

Figure 7.6 One option for an Intranet

Response times are also important and these depend on the way in which protocols and applications are used as much as on raw bandwidth. As already illustrated, the level of security should be appropriate to the value of what is on the Intranet—and different levels are likely to be needed, even in-house.

Once users are put onto an Intranet, there is a need either to register their IP addresses or to use in-house addresses and deal with address translation at the interface to the public Internet.

Rather than carry on with more of these practical design issues, it suffices to say that there are many aspects of an operational Intranet that need to be dealt with but that have not been addressed here. Also, with the focus on data networking there is little common ground with the example in the previous chapter. So, to get a better feel for the issues and options in developing an Intranet, this section describes some established designs that have been put to good use.

In our first example (Figure 7.6), we have a number of local sites connected to a third-party network—BTnet—which is configured to provide a secure link between each node. Interconnection to the public Internet is managed by the third party. Putting this together entails:

- having a DNS Nameserver on the firewall to hide all internal host addresses. Internal machines have access to the full DNS Namespace, whilst the only customer information available to the outside world will be presented by the firewall;

- a mail relay configured to pass main into and out of the customer's internal Intranet segment;

- a news application gateway (running NNTP) configured to act as a relay between an internal news server and a news server based on an external network;

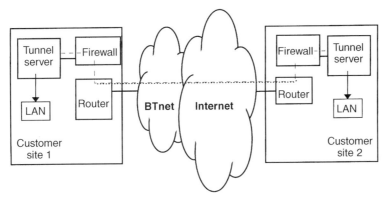

Figure 7.7 An alternative option for an Intranet

- inbound file (ftp) access that can be configured, to some maximum of specified internal hosts and controlled by username/password validation at the firewall, and inbound telnet likewise;

- inbound World Wide Web (http) access attempts rejected at the firewall unless to one of the configured hosts;

- having the firewall configured to synchronise to a nominated external time server, and to provide time services to hosts on the internal segment.

On the outbound side it entails:

- ftp access configured to allow customer hosts to retrieve data from any Intranet or Internet hosts (with sensitive external locations protected by blacklist);

- a proxy WWW server configured on the firewall that permits access to all unprotected Intranet WWW servers and all Internet WWW servers using http. Again, sensitive external locations can be protected by blacklist.

This option works perfectly well and provides users with all of the services they need. Management and security are simplified considerably by having a third party manage the interface to the public Internet. Also, with each local site connecting via a third-party network it is fairly straightforward for each to look after itself. The constraint in the arrangement is that any new sites or rearrangements have to fit in with the third-party network.

In Figure 7.7 we have another option for building an Intranet. On this occasion, the public Internet forms part of the solution. Instead of having one point of contact with the public domain (and hence one firewall to look after), we have a firewall at each separate node in the Intranet. Security is assured through tunnelling.

Tunnelling allows information to be securely passed between a server and another computer over a public network, as if the two were connected by a single physical wire.

The process of sending information through tunnels is simple and straightforward. After authenticating the tunnel client and the tunnel server, information is encrypted by its sender, encapsulated into TCP/IP data packets, and sent across the Internet as unreadable and unrecognisable data.

Once they reach their final destination, the packets are reconstituted and decrypted into a readable form.

In practice, the above picture can be realised using components such as Digital's AltaVista Tunnel, which employs a Public Key Cryptosystem (RSA, after its inventors—Rivest, Shamir and Adleman) for authentication and session key exchange.

The cryptographic identity and keys are tied to the user in this instance, leaving the IP address free to be dynamically assigned. This means that the above picture can be extended to include remote access via the public Internet, or even dial-up access over the public switched telephone network. So the increased complexity on the security side can be offset against the greater flexibility of connection.

As indicated throughout this chapter, the way in which the available standard components are assembled to build an Intranet is what determines the nature of the end product. And there is considerable flexibility to cope with a wide range of needs.

7.7 SUMMARY

In this chapter we have described how an Intranet is assembled. In addition to the physical network design we have explained how and why security and management capabilities need to be built in to a practical Intranet.

The aim throughout has been to illustrate what a Total Area Network looks like from the computing point of view. Hence the focus has been on the way in which the standards developed for the Internet are used to deliver service to a selected group of users.

To effectively address organisational and business realities and deliver the full promise of a Total Area Network, a full-function Intranet suite needs to satisfy three primary requirements:

- efficient individual and group information management (and distribution facilities);

- cost-effective resource management (efficient use of bandwidth and ability to reconfigure according to need);

- administrative control (the ability to configure the network, cope with faults and manage users).

The mechanics behind achieving this set of requirements have been worked through and design obligations and practical issues explained.

REFERENCES

Cheswick, W. and Bellovin, S. (1994) *Firewalls and Internet Security: Repelling the Wily Hacker.* Addison-Wesley, Wokingham.

Frost, A. and Norris, M. (1997) *Exploiting the Internet.* John Wiley & Sons.

Held, G. (1996) *Understanding Data Communications.* SAMS Publishing, Indianapolis.

Norris, M. and Winton, N. (1996) *Energize the Network: Distributed Computing Explained.* Addison Wesley Longman.

West, S. and Norris, M. (1997) *Media Engineering.* John Wiley & Sons.

West, S., Norris, M. and Stockman, S. (1997) *Computer Systems for Glocal Telecommunications.* Chapman & Hall.

8
Network Management

*In the modern world, and still more, so far as can be guessed,
in the world of the near future, important achievement is and
will be almost impossible to an individual if he cannot
dominate some vast organisation*

Bertrand Russell

The information held within computer systems is the life blood of a modern organisation. Without fast, reliable access to this information most organisations would not be able to function for very long. Some companies, such as those providing telephone banking or credit vetting and the utilities such as gas and electricity, are so dependent on this information that they would stop functioning almost immediately if they lost their communication system or even access to their database.

Continual and reliable access to distributed information is now needed. Organisations seek to do business when and where customers require it. They are also looking to maximise the return on costly capital equipment. These drivers are extending the times that the network must be available, so that 9-to-5 operation is no longer universally applicable, especially when networks that cross the globe have no real concept of what 9-to-5 is!

Within individual buildings local area networks (LANs) have been increasingly deployed to provide common access to data to be shared across work groups. These networks are already indispensable to the people they support. As part of a distributed information network, they will become even more central to the operation of many businesses.

As companies rationalise their previously diverse operations, there is a need to control the linking of separate computer systems. This means that useful networks require management support that enables them to provide and maintain common services from a disparate set of components. This capability will become ever more important to the company's core activities, and hence its overall business objectives.

This chapter starts by reviewing the essential problem: that of managing a network of diverse elements to perform to such stringent requirements as those stated above. We build a picture of the general aims that have to be addressed before this can be achieved. We then build on this in considering the practical issues that have to be addressed in managing a real network (whether you do it yourself, or whether a third party does it for you). Finally we outline the fast-maturing network management systems so essential for coping with the volumes of data and speed of reaction required by users.

8.1 TYPICAL NETWORKS

The range of information technology is so wide that no single supplier can deliver a cost-effective solution to every user requirements. Inevitably, real networks are a composite of different piece parts, each from a different vendor. In general, these piece parts work reasonably well with each other to provide the desired range of facilities: mail, telephony, file transfer, etc. For instance, in the networks of large companies, the average number of suppliers is typically between 16 and 20. The end user is often (mercifully) shielded from underlying complexity.

A real network (see Figure 8.1) will involve components from a variety of sources which might well include (NCC 1892):

- private exchanges, local area networks, bridges and routers, multiplexers from network equipment vendors

- telephones, leased circuits and managed data networks from public network operators

- computers, communications software, communications controllers and terminals from computer suppliers.

The challenge facing the network owner is the integration of these separate components into a coherent network in a cost-effective manner (Spooner 1993). The set of components used must not only interwork; it must also be readily operated and maintained as a single resource.

The users of a network, however, are not interested in the individual elements of the network; they are concerned solely with the services provided across it. For them, the network must be managed for effectiveness and reliability so that they get the services they need without being aware of the technical detail of the network.

Diverse networks require systems that enable the network to be managed in terms of the services it supports. It may be possible to manage a piecemeal set of components to perform as required but as the level of connectivity increases, this will become more and more demanding. There will come a point (and this is not far off) when the diversity and complexity of the network preclude anything other than an integrated management approach.

The next section looks at the network as a resource that has to be managed.

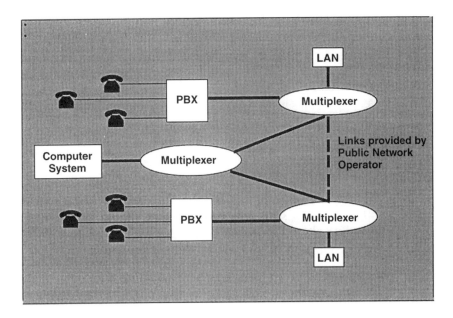

Figure 8.1 Today's typical network

Before going on to this, it is worth dwelling for a short while on what we mean by 'managed'. In many people's eyes, network management is little more than fault detection. A lot more has to be done, though, to satisfy the escalating expectations of users and needs of business. Modern network management needs to cover all activities concerned with operating and maintaining the network to meet the required quality of service and ensure the most efficient use of resources. This covers many different aspects, some of the more important ones are described below.

Service provision

The services that users require should be available on demand. It should be possible to provide and change services with minimal manual intervention. Automated processes should be provided to manipulate network data so that this can be achieved.

Repair

Most of the faults that occur in the network should be identified before they affect service. Network elements should be monitored for faults or degradation of performance. Network performance should be automatically reconfigurable to restore service. Faults should be diagnosed, correlated, prioritised and dispatched for attention as soon as they are found.

Test and performance monitoring

Any network element, anywhere in the network, should be testable from any fault handling point. The quality of service being offered to a user should be monitored in real time. A network being managed by a third party should have a service level agreement (SLA) against which acceptable levels of performance can be measured.

Inventory management

All network elements should be identifiable and their status (e.g. connected, out of service, etc.) and configuration should be known. The network inventory should be accurate and detailed enough to ensure that it is clear exactly what constitutes the network being managed

Order handling

It should be possible to order/reserve resources (e.g. extra bandwidth, new services, etc.) automatically. Checks on available capacity, credit, stocks, etc. should also be available as part of a network management system.

Accounting

Charging, tariffing and credit information should be maintained as part of the management system. This would be partly for information and partly for verification purposes.

There are many items that could be added to the above list, some basic and general requirements such as the security of access, others more specific, such as automated service creation. A picture of the range of what can be done with network management systems is given in Figure 8.2.

It should be noted that not all customers/users will want all of the functions shown in the diagram to be available. The suppliers of network management systems will need to customise their offerings to provide a range of offerings, otherwise cost and complexity may increase needlessly for a user with more straightforward needs. The sheer range of network management is a clear theme of this chapter and can readily be seen to be a driver of the network management tools now on the market.

8.2 THE NETWORK AS A RESOURCE

Organisations are constantly changing their shape, with locations closing, changing size and new ones being added. As organisations focus on their core activities, other activities are contracted out, divested or stopped all together.

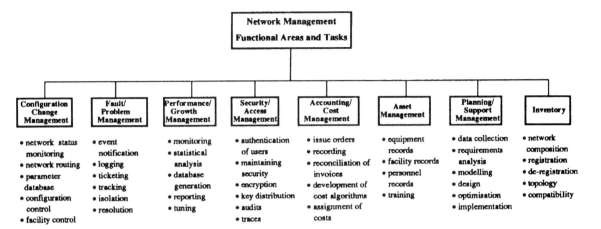

Figure 8.2 The range of functions covered by network management systems

If the network is to support the organisation in this type of environment, it must be flexible.

Traditionally, organisations have used private networks to support their operations. These have the attraction of being reasonably priced while being under the control of the organisation; hence providing greater reliability, functionality and availability than public networks. Since they are fixed assets, private networks usually have spare capacity, paid for but unused most of the time.

Public network operators have responded by offering network services such as managed data network services or virtual private networks that have addressed these concerns and provide high levels of flexibility. With these network services, parts of the large networks owned by the network operator are customised to the individual needs of a particular organisation. The customer thus appears to have a discrete private network.

To be truly effective these network offerings must be capable of being integrated into an organisation's network management structure. Although the network infrastructure may no longer be privately owned, the services it supports must be managed as effectively as though they were.

As companies question the rising cost of operating a complete private network, options such as facilities management, where all or part of the network is contracted out to a third party, are becoming more commonplace.

The options to move between private networks, facilities management and managed network services provide an organisation with a spectrum of solutions from which they can choose the most appropriate for each part of the network. For example, a private network may be appropriate for the UK operation, with a managed data service for the continental European part of the network.

Networks require management systems that enable effective monitoring and control. The key role of the network as an operational resource means that it has to provide an acceptable level of service We now go on to consider the aims and practicalities of real network management.

8.3   AIM 1—EARLY DETECTION

Network management has until recently been reactive. Network managers would wait until the users of the network reported a fault, or a piece of equipment failed and generated an alarm. Upon receipt of the report the managers would investigate the fault and initiate some repair process. This reactive management style results in users suffering loss of service, which is increasingly unacceptable given the key role of the network in supporting the effective operation of the organisation.

Network managers need to monitor continuously the health of the network, identifying trends within the network so that they can take action before services are affected. This proactive management style needs automated tools that can process the large amounts of performance data available from a network. This is especially true in the case of VPNs where a third party is held responsible (and hence penalised) for loss of service.

The only problem with this is that most networks have a significant amount of equipment managed by systems that are proprietary. It may not be cost-justifiable to change these systems or implement standard management interfaces on them immediately. So, gradual migration to an integrated environment (which, in practice, has to be based on standard interfaces) is necessary.

The introduction of network management standards is not uniform across all technologies, though. Newer network technologies such as local area networks or ISDN have quite well developed standards, and others such as PABXs have few.

8.4   AIM 2—CLEARLY DEFINED NETWORK MANAGEMENT CONCEPTS

International standards committees (most notably the OSI) have defined models that describe how to manage real networks. These models use a method known as object orientation, which is designed to cope with the complexity found in today's information systems.

The object-oriented approach decomposes a complex problem into understandable pieces known as objects. The approach looks at a network piece by piece, modelling each separately, so that it is easier to comprehend.

The pieces of the network that are modelled do not have to be physical elements, such as modems or multiplexers; they can also be software elements, for example a file manager on a LAN.

From the management point of view a network resource is called a managed object. Each managed object is described by attributes such as telephone number, address, identity number, etc. Not all the characteristics of a resource may be of interest in management terms. An important issue when modelling a resource is to decide which characteristics are relevant for

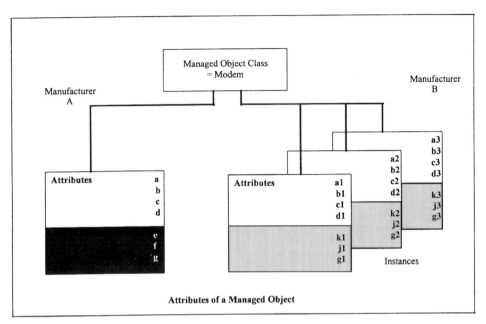

Figure 8.3   Objects and inheritance for network components

network management purposes and which therefore should be defined as attributes. All OSI management standards are based on the concept of manipulating information about managed objects and their attributes.

An individual element of the network is represented by an instance of the managed object. In general there will be a number of instances of an object type which will have common properties. This collection is known as a managed object class.

To illustrate how this works, we can use the example of modems (see Figure 8.3). There is a set of attributes (a, b, c, d) that each instance must have to be classed as a modem. There are other attributes (e, f, g for manufacturer A) which are particular to a given manufacturer's modem which may or may not exist in another manufacturer's product. It is an obvious goal for network operators that the common attributes for an object class are maximised and the manufacturer's unique features are minimised. Of course, for manufacturers, the opposite is usually the case. The extent to which manufacturers will differentiate through features or quality remains to be seen. The former is beginning to give way to the latter.

To produce short, clear and consistent definitions of managed object instances which are free from unnecessary duplication, the concept of Inheritance is used. One class of objects can be defined as an extension of another. All the properties of the original class apply and only new properties need be defined.

To define an object fully, further definitions are required. These are

- operations: the management operations that may be applied to the managed object instance;

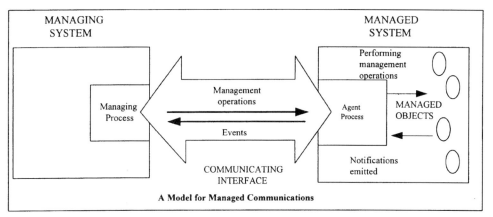

Figure 8.4 Communication in a management system

- notifications: emitted by the managed object instance as a result of an internal event;

- behaviour: the actions of the managed object instance exhibited in response to management operations.

The set of managed objects in a management system is logically stored in a management information base (MIB). This is a conceptual store, which can be physically implemented in a number of ways.

Having defined a set of objects, we now have to devise a means for describing how they interact with each other. This is where the concept of a managed communications model is useful.

Management communications are required when a system wants to gain control or to access information about the behaviour of another, or when one system wants to report an event to another. Each management communication is said to take place between a managing system (containing a managing process), and a managed system (containing an 'agent' process which works on and drives a set of managed objects) as shown in Figure 8.4. Either system may initiate the communication.

It should be noted that this classification into managing and managed systems applies to a single communication. A system may act as a managing system when communication with a 'subordinate' system, and at the same time act as a managed system when reporting events to a 'superior' system. This simple, two party model of communication can be used as a basis for building a variety of very complex multimanager configuration types, such as hierarchical or functionally organised systems.

8.5   AIM 3—NETWORK MANAGEMENT STANDARDS

For interworking between various elements to be possible, network equipment needs to conform to accepted standards. One of the most cherished goals of standards bodies has been to enable heterogeneous networks to be constructed. This noble aim has been slow to arrive, though as suppliers have constantly striven to add value by extending the capabilities of their products, they have added proprietary features including network management. This has led to a multitude of ways of managing individual parts of a network, but little in the way of managing a network as an integrated whole. A consistent set of international standards has yet to allow the management of multivendor networks.

The prime source of standards that do exist for network management is the International Standards Organisation, ISO. This is an agency of the UN in the UNESCO family. Each member country is represented by a single national body, which in the UK is the British Standards Institution (BSI).

ISO standards are being promulgated by the Network Management Forum, a group of major manufacturers collaborating on the development of network management protocols such as CMIS and CMIP. The forum's basic charter is to develop protocols and interfaces that can be used on an industry-wide basis under their OMNIPoint banner.

Network management is open to *de facto* as well as *de jure* standards. There has been significant progress in both areas, but the final shape of standards that apply in practice is far from settled.

Industry-led consortia such as the Open Software Foundation and Distributed Computing Environment are likely to have a significant impact on what network managers use in the long term. For instance, the OSF DME standard (explained in Appendix 2) accommodates both SNMP and CMIP as interfaces on its Consolidated Management Application Programming Interface (CM-API). Several of the more advanced network management systems described later in this chapter plan to intercept DME.[1]

The structure of the OSI network management standards is now described.

Framework standards

These specify the models upon which all other management standards are built. The OSI system management overview (ISO/IEC 10040) outlines the scope of OSI management using four models, as below.

The organisational model This describes how OSI management may be distributed administratively.

The information model This specifies the use of managed objects and provides guidelines for defining managed objects.

The function model This describes network management functional areas. The functional model is a commonly accepted way of describing network management activities. These are

---

[1] A cynic's definition of DME is Distributed Management Eventually.

- fault: identifying and recovering from faults;

- performance: measuring and changing performance parameters;

- configuration: keeping an up-to-date inventory of network resources and controlling changes;

- security: ensuring that the network can be accessed and used only by authorised users;

- accounting: managing the financial aspects of the network, such as charging and asset auditing.

The Communications model This describes how systems will exchange information.

The last three of the above models are backed up and elaborated with standards that allow practical implementation. The organisational model is, however, more fluid, very much driven by user needs. Aspects of this have been touched on already with the 'own-managed or outsourced' choice mentioned earlier in the chapter. The practicalities of network management explained towards the end of this chapter provide much of the remaining information required to decide on appropriate organisation.

Function standards

These specify how to use management communications to achieve a particular objective.

The five areas described in the functional model overlap significantly in practice. For example, the boundary between configuration management and fault management is blurred: repairing a fault may require temporary reconfiguration of the network. In each of the functional areas the activities performed may be similar: for example tests might be performed on both the security and performance aspects of the network. To resolve these issues of overlapping functionality the OSI committees have defined numerous generic functions, such as alarm reporting, event management and test management, which can be applied to any of the functional areas. These are known as function standards and are typified by those listed in Figure 8.2.

Communications standards

These specify the protocols and services that communicate management information.

The common management information service (CMIS) and common management information protocol (CMIP) are a linked pair of standards.

CMIS provides a service definition using commands, or primitives, to perform network management functions. An example of this is the notification service used to report events within the network.

CMIP specifies how to exchange basic management information between open systems.

Therefore in a 'pure' OSI system CMIS commands pass between systems using the CMIP protocol. It should be noted, however, that it is possible to exchange CMIS commands using other protocols. The SNMP protocol is the main alternative, although it is converging with OSI and many vendors support both.

Simple Network Management Protocol The Simple Network Management Protocol (SNMP) has, over a relatively short period of time, become the *de facto* standard for local area network management. SNMP performs analogous functions to the OSI CMIP but has achieved its prominence through a rather different route.

The Internet, introduced in Chapter 2, is controlled by a body called the Internet Activities Board. In 1983 it mandated a suite of protocols known as Transmission Control Protocol/Internet Protocol (TCP/IP) for all data communications using Internet. These days, TCP/IP is available on virtually every LAN installation.

In 1988 a group of scientists responsible for large university and research internetworks created SNMP as a protocol to monitor network performance, detect and analyse network faults, and configure network devices. In April 1989, SNMP became an Internet recommended standard. Since then the standard has been enhanced, primarily by extending the Management Information Base to cover a wide range of LAN elements.

SNMP has been implemented in intelligent network devices ranging from Unix workstations, through bridges and routers to multiplexers, modems and network adapter cards for PCs.

SNMP architecture The SNMP architecture uses four concepts.

- A Manager which runs on a network management station. The manager has the ability to query agents, receive their responses, and set variables using SNMP commands.

- An Agent which runs on a managed network device. The agent stores management data and responds to the manager's requests.

- The Management Information Base (MIB) which is the database of managed objects, accessible to agents and manipulated via SNMP. The MIB assigns names to items according to an OSI registration hierarchy.

- The SNMP Protocol which is an application layer protocol that outlines the structure for communication among network devices.

Information standards

These specify *what* is managed (managed object classes). This group of standards explain the structure of the management information and include an explanation of the object-oriented approach used to model the network. The two important standards in this area are

- management information model: this defines how the management data should be structured, including what managed objects and attributes are;

- guidelines for definition of managed objects (GDMO): this standard tells implementors how a specific object definition should look. The goal of GDMO is to ensure consistency and completeness of object specifications.

In these standards the management information base (MIB) is the conceptual repository for management information. The actual set of object definitions is held in a management information library (MIL). However, the term MIB is now widely (mis)used to refer to both concepts.

The above standards go a long way to allowing network management products to interwork, at least in theory. In practice, though, more is required. Those who build and operate interoperable systems must agree on exactly how the standards are to be implemented.

These agreements have come to be made through implementor's workshops and are documented in international standard profiles (ISPs). These define how a combination of bases standards should be used for a given application. In addition, a profile specifies restrictions on values of parameters, the choice of permitted options and all those other practicalities required to ensure that systems can be tested for interworking.

Once all of the abstract details how to manage a network are sorted out, there remains a further management challenge, that of applying the available knowledge, tools and techniques. We now move on to some of the more important issues that have to be addressed within an organisation to acquire the level of management appropriate for their network. The term 'management' now denotes action rather than concepts.

8.6 PRACTICALITY 1—A NETWORK STRATEGY

As indicated at the start of this chapter, networks are increasingly made up of a diverse range of products and services, and the best blend for an organisation will change over time. It is the network that benefits the organisation, an increasingly vital resource. Thus, network management is a key enabler for optimising the investment (Valovic 1992).

One of the key steps is therefore the creation of a network strategy. This provides the basis from which a network management plan can be derived. It should include the main network upgrades and the strategy to be adopted in meeting business needs. The network strategy must take account of viable network options within the planning horizon. For example, will network services be a better option than continued investment in private networks?

Network technologies should not be excluded from consideration because they do not readily fit a particular network management structure. The opposite should hold: the network management should be arranged to enable the most appropriate network technology to be deployed, to a required level of quality of service (Held 1992).

Inevitably, some parts of a real managed network may be accessible via standard interfaces and protocols, others may not. In practice, part of the strategy must address not just what is managed but also how each element is accessed and what level of control can be exercised over it (e.g. can an element be automatically enrolled when it is added to the network?). Typically, newer equipment (e.g. Frame Relay switches) will come with interfaces that permit effective remote management. Older network elements will have less (or indeed, no) remote management flexibility.

A consequence of this is that procedures for management need to be developed to suit this (inevitable) mixture. There is little point in having a centralised management facility unless you can access and control most (if not all) of the network components. Alternatively, it would be an opportunity spurned to use a variety of local and/or proprietary management systems for a network that could be looked after by an integrated management system. The tiers of sophistication in network management are revisited in a subsequent section dealing with automation.

8.7 PRACTICALITY 2—MANAGEMENT ORGANISATION STRUCTURE

There are a number of ways that a network management operation can be organised. It can be organised geographically, by network equipment type or by the services offered. There may be one single point of contact running a user-facing help desk or a variety of contact points depending upon the services required by the user. A business analysis of the activities to be performed will help in understanding and deciding the best organisation structure to adopt.

The management models in the OSI framework standards and the Forum user requirements documents will help to ensure that the scope of such an analysis is complete. The function standards will help in the detailed consideration of particular functions the organisation must perform.

Further to this, the management of the network may be part of an organisations core business or it may be a service that can be contracted out to a third party. This latter option, commonly referred to as outsourcing, is becoming increasingly attractive as the option of remotely controlling a set of disparate network components becomes more viable.

Whether outsourced or in-house, roles, responsibilities and levels of performance all need to be clearly defined. It should not be forgotten that the network is a resource.

8.8 PRACTICALITY 3—A MANAGEMENT SYSTEM STRATEGY

Few companies have sufficient resources to build management systems from scratch, when there are prime business applications often competing for the same resources. Thus, network management applications are predominantly purchased and adapted to organisations' needs.

What each user requires is a strategy for the deployment of network management systems based on their particular organisational needs and plans. Operational constraints, such as the numbers of people employed to run the network or budget allocated to having it managed, will impact on the choice of systems. Another important factor is the planned evolution of the network; if it is to support an increasingly distributed user base, then the management system should have the flexibility to grow with the network it is looking after.

In either case, the management system strategy must identify the functions that the system must perform and the network domains it will cover. From these, the system requirements may be defined and a policy on the use of open standards derived.

Most architectural models for the deployment of network management have some systems dedicated to a particular domain (such as modems or LANs) and some general purpose systems for the whole network (such as a help desk or billing system). The architectures allow choice of a protocol (such as SNMP or an OSI protocol for LANs) for each domain depending on individual circumstances. The interworking between management systems

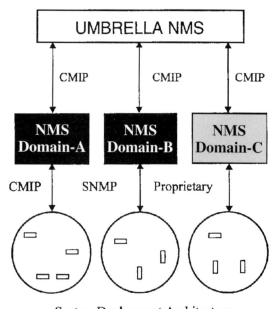

System Deployment Architecture

Figure 8.5 A general view of network management

and the general purpose applications is always based on OSI standards.

Figure 8.5 gives a generalised view of these architectural models. By using such a model as part of the management system strategy it is possible to identify key interfaces and required functions. A considerable amount of thought has been applied to management strategy over the years with, for instance, the Telecommunications Managed Network (TMN) publications providing the basis for many implementations.

A clear strategy can then cater for the adoption of system interfaces and applications software that implements standardised functions.

8.9 PRACTICALITY 4—AUTOMATION

Effective management of real (i.e. complex) networks really requires automated tool support, if management strategies and targets are to be met. Over the last ten years, a whole host of management tools have appeared. The majority are proprietary and allow a limited amount of control over specific parts of a network (e.g. LAN elements). The importance of automated management systems is one very clearly taken by many of the virtual private network providers, who have invested heavily over the last few years in integrated network management systems (Helleur and Milway 1991).

Table 8.1 summarises the various levels of capability in current network management offerings.

There are many facilities in the 'element manager' category on the market place. These are commonplace tools and are routinely used by many systems managers as part of their LAN management operations. Domain management tools are also fairly widely deployed, especially within large companies, to look after computing infrastructure. For the end to end management of an organisation's network, there are fewer examples of the use of automation.

Table 8.1 Hierarchical structure of network management facilities

Tier 1 *Element managers*
Basic systems that monitor devices such as multiplexors, PBXs and other network components (typified by any of the proprietary management systems that can be purchased to monitor/control the basic hardware and software)

Tier 2 *Domain managers*
The next level in the hierarchy. Can be described as multi-equipment systems that integrate information from various element managers (typified by management tools such as Sunnet Manager and SynOptic's LattisNet)

Tier 3 *Integrated network management systems*
These are the powerful systems that collect and aggregate information from a wide range of sources to enable centralised management of complex and diverse networks. They are typified by IBM's NetView and BT's Major Customer Service System.

This is not really surprising, as only a few tools have been developed that cope with the full set of network management functions explained earlier. These tend to be extremely powerful, standards conformant, systems that allow a wide range of LAN and WAN elements to be managed.

Since the future of networks is likely to be critically dependent on effective management, we close this chapter with a brief history and description of some of these tools, namely

- IBM's NetView
- AT&T's Accumaster
- BT's ServiceView

These systems are selected for illustrative purposes, and readers should not consider the inclusion of a vendor's product in this section as any sort of endorsement. Similarly, the omission of a product should simply be viewed as the author's perspective, since space constraints do not permit an examination of the multitude of commercially available systems. Also, given the fast moving world of network management, the descriptions below should be viewed as snapshots that are designed to illustrate how management systems can be organised.

NetView

Until 1986 IBM marketed a series of separate network management products, including Network Communications Control Facility (NCCF), Network Problem Determination Application (NPDA), Network Logical Data Manager (NLDM), Network Management Productivity Facility (NMPF), and VTAM Node Control Application (VNCA). In 1986 IBM announced its SNA Management Series (SNA/MS) architecture as a blueprint for its network management strategy. This entailed describing the functions and services required for the management and control of SNA network components. The philosophy was to provide a 'single view of the network'. The first IBM product which implemented this was NetView, which can be viewed as the consolidation and integration of that vendor's previously mentioned network management products.

NetView provides a centralised, integrated management capability for SNA networks, as well as a link for the management of non-SNA products, while providing a consistent operator interface. In addition, with an appropriate password any authorised user can manage and control an organisation's network from any SNA terminal in the network.

As an application on a host processor, NetView is designed to be the focal point of all SNA network management elements. In the NetView world, the focal point is one of three defined types of network elements, the other two being entry and service points.

The focal point, as its name implies, provides a centralised set of network management functions and consolidates data received from network elements, in effect functioning as a repository.

Entry points serve as SNA control locations that provide the focal point with network management data about itself, as well as devices connected to the entry point. An example of an entry point would be an IBM 3174 control unit which, under the control of NetView, can provide information concerning itself, as well as terminals and printers connected to it.

The service point is the third NetView element. It is designed to provide a window to non-SNA products. It receives network management data from non-SNA devices, translates the data into a standard SNA data management format called network management vector transports (NMVT) and routes the translated data to a network focal point. NetView/PC which is an IBM software package designed for operation on IBM and compatible personal computers, is an example of a service point.

Accumaster

Hard on the heels of IBM came a number of other suppliers, all striving for the holy grail of integrated network management. One of the most significant was AT&T who, in 1987, announced their Unified Network Management Architecture (UNMA). This was intended to be a framework for developing integrated, end-to-end voice and data management in a multivendor environment. Under UNMA, users can obtain a variety of network control capabilities, ranging from on-premises network control to contract management performed by AT&T at various vendor network control centres.

The UNMA hierarchy can be viewed as an inverted tree structure (see Figure 8.6). At the top of the structure are network elements, such as modems, multiplexors, LANs, and other types of customer premises equipment. This equipment, which can represent products from many vendors, will use proprietary network management protocols unique to the vendor's network management system. Each of those network systems, which may represent a processing box rather than a computer with operator console, is defined as an element management system by AT&T. Each element management system will pass its information to a central site for processing under the UNMA structure, denoted by the box labelled Integrated Network Management System in Figure 8.6.

The key to the ability of third party vendor equipment to operate under AT&T's UNMA was their ability to support a set of communications protocols called the Network Management Protocol (NMP). Although based on the OSI structure, the NMP's network management functions did not represent existing standards; work was still in progress in this area when AT&T developed its approach. In retrospect this had a significant bearing on the commercial fate of Accumaster.

UNMA was enabled in practice by the Accumaster family of network

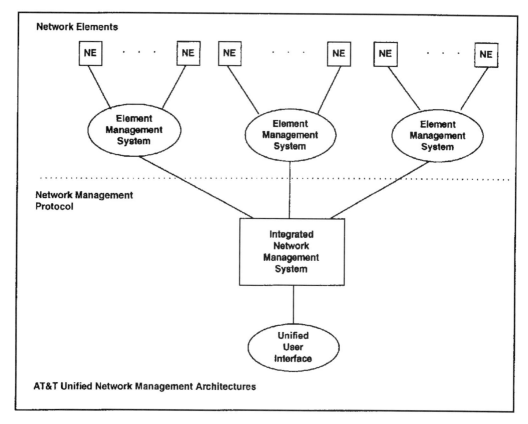

Figure 8.6 Structure of Accumaster

management tools. There were three distinct ways of accessing the Accumaster facilities: a standalone terminal or workstation, an Accumaster Services Workstation, or the Accumaster Integrator. The Services Workstation was designed for organisations that needed to control multiple AT&T network services through one workstation. In comparison, the Integrator permits organisations to manage their own network services, as well as their on-premises equipment, literally providing an end-to-end network management capability.

In operation AT&T offered a service to users to have their network managed at an AT&T Customer Service Centre or at their facility. If the network was to be managed on AT&T premises, that vendor would staff the network management system. In addition, AT&T offered organisations personnel to perform the work on a client's site. Both these options are commonly referred to as outsourcing or facilities management.

As might be expected, AT&T offered users a broad range of network management services to monitor and control communications facilities obtained from that vendor. For instance, by mid-1990, there were more than

20 network management services available from AT&T. Even so, the challenge of integrated network management caught up with Accumaster a few years later and the product was withdrawn; a victim, perhaps, of moving into a complex area before the base standards were in place.

ServiceView

BT's entry into the integrated network management arena came a couple of years after the appearance of the early trailblazers. In many ways, this has proved to be a valuable ploy, as the ServiceView product was designed to operate to a reasonably stable set of concepts and standards.

ServiceView (originally known as Concert) was developed very much in line with current and emerging standards; the OSI standards outlined earlier and the Omnipoint (industry-led) standards developed by the Network Management Forum.

To emphasise the need for standards, BT provided assistance to its suppliers to develop compatible element managers, through documentation on ServiceView interfaces, and a complete ServiceView interface toolkit. This included software source code and supporting documentation designed to assist in the implementation of the CMIS functionality and protocols, and managed object classes compatible with the Network Management Forum definitions and any extensions supported by the ServiceView system.

BT also supplied a set of simulation software which could be used to pre-test element manager interoperability with ServiceView prior to a final interoperability test which will be conducted by BT. The partners programme, launched during initial product development, promoted this approach. It had dozens of members, including nearly all the major network equipment manufacturers.

In terms of structure, ServiceView follows the same sort of approach as Accumaster with a hierarchy leading from element managers (which look after a variety of proprietary components) to an integrator (which provides core facilities). In terms of functions provided, ServiceView is probably among the most powerful tools for total network management. It allows all the basic facilities described earlier: fault detection and collation, inventory configuration and control, automated order facilities, etc.

In addition, ServiceView features a number of mechanisms to ease the management of complex systems. One example of this is the scoreboard facility which provides a summary view (based on linked icons, colours and counts) of associated objects on screen. This allows an operator to keep track of the status of a complex set of transactions without getting lost in detail.

In common with other network management tools in its class, ServiceView is not one tool but a whole product set. From the user point of view, there is a series of management features; these are summarised in Figure 8.7.

In practice there is significant similarity of look and feel across this range. In

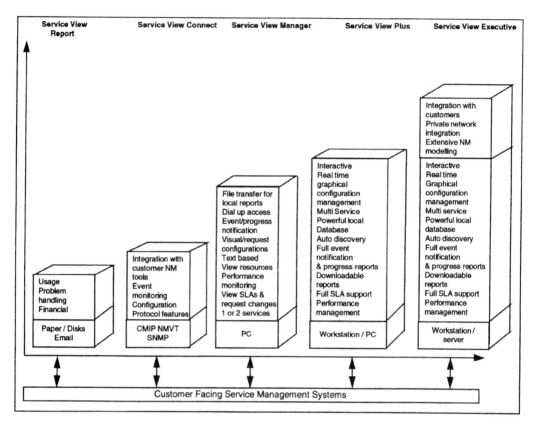

Figure 8.7 The range of ServiceView offerings

the above diagram, the range starts with basic information, provided via a third party, through PC access to a range of passive management facilities, right up to complete virtual network management capability provided through a workstation/PC. The ServiceView executive shown in the diagram has been used to manage large complex networks that span the globe.

To complete this section, a view of integrated network management from the user's point of view. Systems like NetView and ServiceView all allow the aim of virtual private networking to be realised; Figure 8.8 gives some idea of what this will actually look like.

In this diagram the switches, modems, leased lines, etc., report via their own management systems. The value added by the integrated network management tool is to integrate all this information and show it in a way that is relevant to the user (i.e. in relation to their network). The capability of the tools described here allows for flexibility, with the user able to arrange things in terms of what is required rather than in terms of concrete technicalities.

In effect, network management systems are evolving to enable management at a service level; indeed the more recent developments in systems, such as

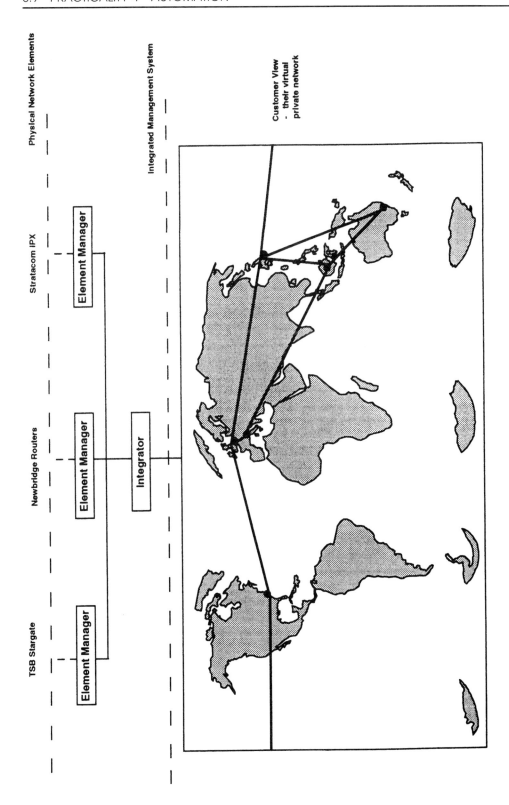

Figure 8.8   Physical and logical network views

ServiceView, aim to allow control over network components in terms of the services being rendered to the end users.

The day of the practical VPN is nigh and it seems likely that increasingly powerful network and service management systems will enable the flexibility that will be required in the information age. The ability to view a worldwide network (as shown in Figure 8.8) and to control effectively the services it provides from a remote location is by no means a pipe dream (BCS 1994).

8.10 SUMMARY

The key message of this chapter has been that the power of modern networks will only be harnessed effectively if they are properly managed. Given that networks are now a vital resource that enable organisations to compete, the ability to effectively manage a network will, increasingly, differentiate the best from the rest.

The still rather nebulous term 'network management' is defined and explained here in the context of what a typical network looks like and what a network manager has to do to keep it in line with the users expectations. We emphasise that there is a lot more to this than simple fault detection and correction. The flexibility, reliability and quality of service required to take full advantage or a modern network means that the network manager must be able to pre-empt failures, dynamically reconfigure bandwidth and services, and a whole host of other things.

Following on from this general picture of network management, we explain the basic concepts and the emerging standards that embody them. Finally we outline some of the important practical issues in managing real networks, including an overview of some of the powerful tools that will enable a disparate set of lines, terminals, servers and other components to be operated as a single entity. The final message is that virtual private networks are coming of age as a practical proposition, and that network management is the enabler of commodity communication services.

REFERENCES

General

BCS Medallists report (1994) The global office. *Computer Bulletin*, February.
Held, G. (1992) *Network Management—Techniques, Tools and Systems*. John Wiley & Sons.
Helleur, J. and Milway, N. (1991) Network management systems; introductory overview. *BT Telecommunications Engineering Journal*, October.

NCC (1982) *Handbook of Data Communications.* NCC Publications.

Spooner, M. (1993) Network management in a modern digital environment. *BT Telecommunications Engineering: Structured Information Programme,* Chapter 13, Unit 3.

Valovic, T. (1992) Network management: a progress report. *Telecommunications,* August, 23–32.

West, S., Norris, M. and Stockman, S. (1997) *Computer Systems for Global Telecommunications.* Chapman & Hall.

OSI network management standards

FRAMEWORK STANDARDS
ISO/IEC 10040 OSI System Management Overview
ISO/IEC 7498/4 OSI Management Framework

FUNCTION STANDARDS

ISO/IEC 10164-1 Object Management Function
ISO/IEC 10164-2 State Management Function
ISO/IEC 10164-3 Attributes for Representing Relationships
ISO/IEC 10164-4 Alarm Reporting Function
ISO/IEC 10164-5 Event Report Management Function
ISO/IEC 10164-6 Log Control Function
ISO/IEC 10164-7 Security Alarm Report Function

COMMUNICATION STANDARDS

ISO/IEC 9595 Common Management Information Service (CMIS)
ISO/IEC 9596 Common Management Information Protocol (CMIP)

INFORMATION STANDARDS

ISO/IEC 10165-1 Management Information Model
ISO/IEC 10165-2 Definition of Management Information
ISO/IEC 10165-4 Guidelines for Definition of Managed Objects
* Note DIS 10165-3 was merged with -2 and was not issued.

SNMP standards

RFC 1155 Structure and Identification of Management Information for TCP/IP based Internets
RFC 1156 Management Information Base for Network Management of TCP/IP based Internets
RFC 1157 Simple Network Management Protocol
RFC 1158 Management Information Base for Network Management of

TCP/IP based Internet: MIB-II
RFC 1230 IEEE 802.4 Token Bus MIB
RFC 1231 IEEE 802.5 Token Ring MIB
RFC 1243 AppleTalk MIB

9

Survival in the Information Jungle

People always overestimate what will be possible in the next two years . . . and underestimate what will be possible in the next ten

H. G. Wells

Anyone reaching this part of the book will have picked up a lot of information about new network technologies and the role they promise to play in the information age. There are yet more facts and detailed analyses to follow in the appendixes. But what are the key messages that emerge? To complete our story of the information revolution, let us consider what it all means to you.

This is not an easy brief. Even with the benefit of hindsight it by no means straightforward to make sense of the current situation. Predicting future evolution is inherently uncertain and something that we will dwell on only briefly. Ideas of teleprescence and virtual reality make for intriguing reading and beguiling pictures but they do not help much in planning the evolution of a network. Bearing in mind the need to make sense of the information already available and that the future will not be forged by technology alone, the final part of this book is about converging not diverging ideas. We now aim to focus on the road ahead rather than the scenery along the way.

9.1 THE SHAPE OF THINGS TO COME

So far, we have tried to concentrate on fact and to put that fact into context. This is not enough on its own, though. A catalogue of what is possible now is of little value and some sort of informed prediction is necessary, especially in such a fast moving area. One of the most difficult aspects of planning is ensuring that items are available when they are needed, even though that need is some way down the track. The predictive element of this book has veered towards the low risk, but all prediction carries some uncertainty

It is possible, for instance, that flaws will be found in ATM as it is pressed to deliver new and faster services. Perhaps the real-time demands of telephony will be too much. But in all probability this will not be the case. In the long term the likeliest scenario is seamless, multiservice, total area networking, based on ATM.

The real issue is how to get there; and that all depends on where you are starting from, who you are competing with and a whole host of other factors. There are as many roads to total area networking as there are local circumstances. It would be inappropriate (and probably impossible) to give any sort of general prescription for an optimum way forward. The core of the book has tried to paint an unbiased picture of the main technical components and organisational influences. These will all have some sort of impact; the key will be to put them into local context and to effect plans that match local circumstance.

There are a number of general issues that should come to the forefront for nearly everyone—the ten safe bets for the future. The greatest overall impact in the information revolution will be down to the following factors.

- Speed of technological change, impact of legacy systems and a host of social and business factors mean that prediction will always be suspect. Survival in this jungle will be down to broad understanding and awareness backed up with detailed, focused planning for specific circumstances. There will be no panaceas or silver bullets.

- The breadth of the technology that enables the information age means that the traditional specialisms of telecommunications, computing and business analysis will be required in one package. A compartmentalised view will inhibit, not enable, total area networking.

- New applications will arrive in the information market with increasing rapidity. The dynamics of total area networking will follow the dynamics of the computing industry and this will increasingly dominate the pace of developments in telecommunications. The successful players (and this covers both the enterprises and the suppliers) will be those who balance quality (both of planning and of product) with speed of reaction.

- There will be yet more technological solutions (although there is little evidence in the industry at present of thinking beyond ATM). The key will be to focus on people's work patterns. These are more stable than technology and (as outlined in Chapter 2) provide a moderating influence on the introduction of new services and systems.

- A corollary of this is that the winners will be those who have the culture to encourage/promote/support/mandate the adoption of organisational structures and working practices that capitalise on total area networking.

We have explained what is available across the board but different organisations will require different blends. Informed planning is key.

- Those who dilute core business by dabbling in an increasingly complex information support arena will be disadvantaged. Sure enough, some in-house technical expertise will have to be maintained but this is likely to be increasingly a planning, control and evaluation function. Outsourcing management will be but one of the emerging skills of the Information Age.

- The future is in multimedia: text, graphics, voice, video. Total area networking will have to provide the platform for this to be delivered as a seamless service, not a collection of independent overlays and derived services. Multimedia applications imply a multiservice network to carry them on: and to have this you need a scalable infrastructure. ATM is the current best base, intranets the best option.

- Organisations will shrink and make increasing use of third parties to supply specialist, non-core services. This means that it will become increasingly important to manage customer—supplier relationships. The interdependence of businesses will make it vital to assess, select and guide appropriate partners.

- Where you don't need the latest technology, don't pay for it—the ISDN will be a natural choice for the economy seeker. The leading edge of total are networking is likely to be expensive. There should be considerable competitive advantage on offer to justify that expense.

- Globalisation will drive the need for distributed working which, in turn, will promote the need for multimedia (as a suitable support for complex human interaction) and the increased use of multimedia will require increasing amounts of bandwidth.

So, the Information Age will have at its core a multiservice network that enables users to communicate using voice, pictures, text and video. From the user organisation's point of view, this network will be a resource that can be configured to meet its needs.

The 'when' and 'how' is debatable, but a reasonable guess if that ATM connected LANs will be in place by the late 1990s, with total area networks following within two or three years.

The quote that introduces this chapter has been proved over and again in respect of technical advances. The 'safe bets' listed here pose a challenge to both the suppliers and to the end users, those organisations whose future profitability lies in generating, finding and processing information. The former will have the opportunity to sell connections, management capability and relevant applications. Theirs will be a technology-led market. The challenge for the latter will be how best to exploit a new capability, to plan their investment in networks to achieve payback. Some suppliers may also be end users (and vice versa). In either case, though, the next part of exploring the road ahead is to see what sort of tools and instruments are available for navigation.

9.2 HORSES FOR COURSES

The main purpose of this book has been to consider a range of technical issues in the context of people's everyday work. In doing this we have had to bridge a number of specialist areas and resolve the different approaches and terminology, especially at the points of overlap. This has not been easy, even with decades of experience in computing and telecommunications, the authors have (occasionally and for brief periods) found themselves a little lost. There is a serious point here: that non-specialists will become confused too. Worse still, they will be sold something that turns out to be a millstone that loses them business, ties them to a particular vendor, costs them dear to maintain or simply does not solve the problem it was bought for.

Given that a perfect understanding of the world of information networks and applications will always be preserve of an exclusive few, then it becomes important to give more general access to a vital area. An understanding of relevant technology is but one part of this, necessary but not sufficient.

The Information Age will be functionally rich, a positive jungle of technology, applications and services; and the jungle is neutral. It is unlikely that things will evolve the way you would like them to; those who make capital in the Information Age will do so through their own initiative.

We now outline a set of tools that allow sensible decisions, plans, etc. without technical omniscience. Not a silver bullet, but an initial analysis that, with a reasonable degree of technical understanding (and having got this far, the authors sincerely hope that this can be assumed) puts you in the driving seat.

To do this requires a mixture of planning and understanding: knowing where you are, where you want to go and what is available to effect the change. Deciding where and how to invest in a raft of technology that, on the face of it, enables virtually anything, is not easy. It relies as much on inspiration, experience and judgement as it does on procedure and technique. Even so there are ways of improving your hit rate and a range of sensible steps can be taken. A first step in this is to recognise that the assessing information is like judging beauty. It is very much in the eye of the beholder and factors such as perishability, exclusivity and accuracy all have critical impact.

In the light of this caveat, rather than try to put absolute figures on the value of information, this section outlines some of the key factors that should be considered. These range from some straightforward financial factors such as the income generated, through to very nebulous factors such as the goodwill earned. These examples typify two extremes.

There is commercial investment, where the aim is to enable a job to be done or for something to be developed so that it can be sold (e.g. buying Frame Relay to interconnect two sites). This category is relatively straightforward to cost, as you are either earning or saving money, time, etc. This does not, however, guarantee a complete answer, as other factors such as development, training, installation costs and likely maintenance expenditure are difficult to determine precisely.

There is strategic investment that supports an essential function within an organisation. The value of such investments often bear little relation to the cost of purchase or development. As such, the assessment in this category is inevitably highly subjective.

Other categories could be added to the above list but the inherent difficulties in appraising investment in new technology are adequately illustrated. Even then, assigning value is not the end of the story. Value is generally far from static: that which is priceless today may be worthless in a couple of years, months, even weeks. The converse is also true. To complete the picture, some notion of the predicted value must be added.

It is helpful to put the above in context by considering very briefly the life and death of an old (but typical) information system. The original specification for this was produced in 1969 in the form of typewritten paper document. Although somewhat dog-eared, this document is still available today; it still has some value. The software from the specification was produced (in the form of a FORTRAN program, stored as a set of machine-readable punched cards) and this still runs efficiently. The PDP11 platform (the original machine purchased for development) is still the main processor to which users connect for this application. After many years of faithful service, though, one request to add a new facility caused a few (readily predicted) problems.

- There is no card reader on the computer; the last one was seen in the Science Museum a few years ago.

- There is no FORTRAN compiler which handles the dialect in which the program was originally written.

- The platform is becoming increasingly prone to faults. The one person in the area who understands the machine is due to retire next year. In any case, their stock of parts is nearly exhausted.

- It is proving increasingly difficult to interface the PDP11 to other systems. The cost of the interface units is becoming comparable to the cost of a replacement system.

These problems, to which others could be added, typify the lack of stability of the support environment for any piece of software, and hence for any modern information-based application. This flux in systems has important consequences on the persistence of information. The last 20 years have seen more information lost because it could not be interpreted than has been lost through flood, fire and war since printing was invented. A combination of distribution and network intelligence will make this possibility more acute.

What to do in countering this sort of circumstance depends on a number of

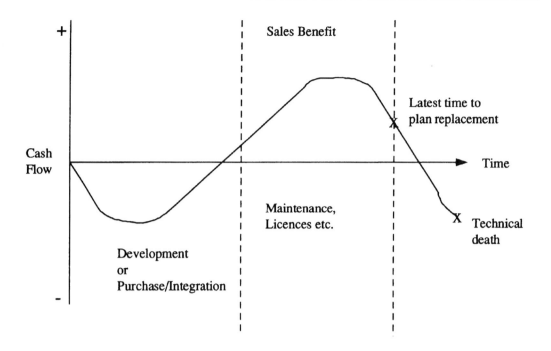

Figure 9.1   A whole-life costing model

factors: the cost and risk of resurrecting the information or code, revising it so that it can reside on a new platform, the cost of losing the information, etc. The point here is that it will become more and more important to identify and manage key processes and information so that they too recoup (at least) the cost of their initial purchase or development.

   If all of the value factors in this could be quantified, it would possible to derive a picture of cost/benefit from inception to death. Then it could be claimed that an organisation's information is managed as an asset and that it is acquired, used and, eventually disposed of on a commercial basis.

   However, this is not really viable given the complexity and pace of development apparent throughout this book. As a basis for planning, some form of model is needed to help balance maintenance to a given standard of quality (which costs money) against the benefits that accrue. This general requirement is common to any modern information-handling application or network investment (most of which are predominated by their software content). Figure 9.1 illustrates a whole life cost model (originally called the software death cycle) developed to meet this requirement.

   The above figure shows the initial cost of development (or purchase) as pure expenditure (i.e. time and/or money taken to produce the product with no income) and the net earnings are assessed from direct sales income minus the cost of support. In practice, the situation can be fairly complex: factors such as the number in use, the build and distribution costs, the implementation cost and the sales and maintenance income all had to be assessed before the

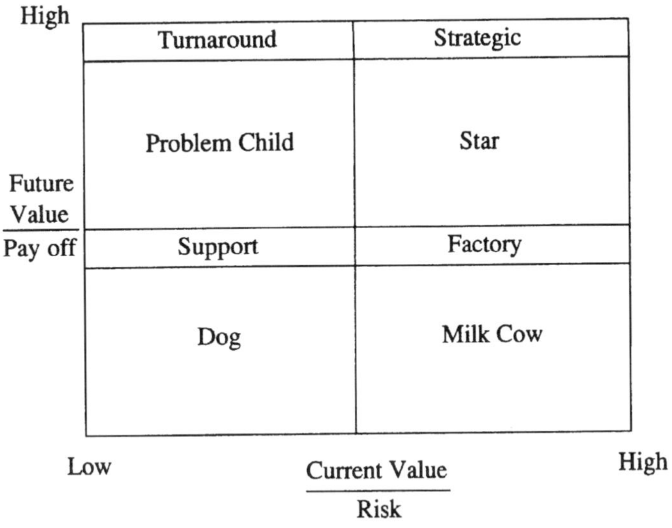

Figure 9.2   A value system to assess a system

overall picture of cash flow for this particular system could be reasonably drawn up. Also, the scaling of the curves in the model has to be determined through audit and assessment.

At some point, the cost of ownership begins to outweigh the benefit of its continued existence and this is the point at which net income becomes negative. This is the point at which the system should be retired from use. In an ideal world, the replacement system (assuming that its function is still required) should have been planned before this point.

Once a system or service has been identified and a value has been assigned to it, a rational plan of what to do with it can be drawn up. This involves discrimination of a different types of investment; planning a balanced portfolio of applications, rather than assessing the viability of one in isolation. A mechanism for such planning is illustrated in Figure 9.2.

There are a number of strategic planning and assessment tools that follow the format of Figure 9.2. This reflects the fact that no one viewpoint can yield all the answers; a number of analyses are required, each of which allows some aspect to be considered objectively. For the purposes of the analysis illustrated in the figure, two key drivers are relevant: present and predicted values. Both these values have to be assessed within the user terms of reference. The point of considering these two values is that they have a direct bearing on what should be done with a particular system. This is best explained by describing the details of each of the points shown in the diagram.

Experimental groupworking ('problem child')

This is a potentially powerful new approach. At present it has only limited application but this is likely to change with advances in technology. The support system is therefore of limited current use but has the potential to become very valuable. The purchase/development strategy is 'turnaround'—rapid deployment of trial technology.

The strategy adopted for systems falling into this category is to build on their potential by investing in their development. This implies careful control over the variants of the system which evolve so that options are not closed prematurely.

Essential network services ('star')

This category contains those application that are vitally important now and are likely to remain so for the foreseeable future (e.g. electronic mail, file transfer). These services and associated systems are essential to the effective operation of the business, and they cannot be readily replaced.

The strategy adopted in this category is to ensure maintenance and care. For support systems, this means regular back up of data, system changes kept under strict configuration control, etc.

A general-purpose application ('milk cow')

The value of this service/system is high, as it is currently used to support an important function. This category is exemplified by a spreadsheet that holds the current budget figures for the department. In the longer term another spreadsheet on a faster machine (on the other side of the world) will do the job more effectively, so the long-term value of the program is low.

The strategy adopted for systems falling into this category is to harvest the benefit of their use. Little effort would be spent on improvement as their usefulness must be milked for all it is worth before they are superseded.

A non-standard accounting system ('dog')

This system has no particular value at present. It was put into use at one location just to see how it would perform. The problems of interfacing it into the rest of the organisations network would be significant. It is unlikely that it will have any long-term value, even if functionally sound.

The strategy adopted for systems falling into this category is to remove them!

The development of a system valuation scheme such as it described above is only the first part of treating information and the systems put in place to access, store and process it as an important asset. How you decide on how to populate the four areas in Figure 9.2 and manage movement between them remains as a challenge best met by an informed reader. The key point, though, is that the complexities of technology do not permit effective management at a detailed level alone. Abstractions that highlight business fit, development strategy, supply tactics, etc., will have to be applied to minimise inevitable risk.

9.3 SURVIVAL IN THE INFORMATION JUNGLE

The quotation at the start of the chapter has proved to be particularly true of information technology. Fads may come and go on a yearly basis but the world of the information worker is unrecognisable as that of the late 1980s. The speed of change may seem slow day by day but the effect of new technology through most people's working life has proved enormous. This is unlikely to abate and will, if anything, accelerate.

Those who live in and push back the frontiers of the information jungle are the modern day explorers who dare to tread into the unknown. Like their predecessors, they will require a survival kit; this jungle is neutral and survival is optional. Abstracting from the many trends, drivers, technologies, etc., here, the most important elements in the information jungle survival kit are likely to be as follows.

Managed, not just controlled

An information-intensive organisation is a dynamic entity that needs to be managed, not a stable operation that simply needs keeping under control. Just because information, expertise, etc., is distributed and intangible doesn't mean that it should not be treated as a valuable resource. For some people the transition from the tight control over a tangible local unit to the management of a loosely couple partially autonomous one can be painful. This will be an essential aspect of the Information Age, though the ability to manage at a distance will be a prime skill.

Understand colliding technology

The move from analogue to digital switching of telephony calls along with the drive to connect computers sparked an inevitable collision between telecommunications and computing. With increasing network intelligence being provided by computer-controlled switches, the distinction (at least at a technical level) between distributed computing and telecommunications is becoming very blurred. Both rely on software to control the services and application they provide, both are driven by the doubling of available computer power and memory every other year. Despite this convergence, though, there remains a significant diversity of approach which needs to be bridged. A good understanding of both sides is a prerequisite to capitalising on what is on offer. Conversely, a lack of awareness (of either telecommunications or distributed computing) is a recipe for disaster

Capitalise on global strength, local presence

Distribution of an organisation has a number of aspects. With different places being governed by different laws, tax arrangements, business conditions and working practices, there is always the prospect of turning them to advantage (some countries allow generous tax concessions against R&D work). In addition to this, there is market and people flexibility. Organisations based in one culture or location can suffer 'groupthink'—the shared ability not to solve a problem or exploit an opportunity. The extension of a company culture to draw on a wide range of approaches is not easy, but it provides intellectual as well as market diversity.

Technology moves faster than people

At the end of the day, technology only adds value when people put it to work. In general, people tend to change their practices somewhat slower than technology changes. It will become increasingly important to revise structures to capitalise on faster, better ways of doing business. Rigid organisations, governed by an inflexible quality management system will be left behind in the Information Age. They may not die but they will be relegated to the commodity, low margin, economy end of the market. Those involved in planning and operating the value seekers will need to ensure that people are trained, informed and organised for the information age, not just bolted on as an afterthought to new technology.

9.4 INTO CYBERSPACE

The telecommunications network that currently spans the globe is the largest human construction of all time. This year it will carry over 1000 000 000 000 telephone calls, as well as a significant and growing amount of data, between every country in the world. Virtually everyone now knows how to use a telephone; the complexity of a huge machine is masked behind a simple, intuitive and commonly accepted user interface.

The information network is nowhere near this level of sophistication, in terms of being part of the fabric of society, but this will change, and fast.

Perhaps the list of technical advances presented as Figure 2.1 in Chapter 2 will read like Table 9.1 by the time that we are ten years into the third millennium.

The last of these supposed events is interesting in that a newspaper is

Table 9.1

1999	SmartCards widely used to gain access to network-based services
2001	Portable devices for access to network service commonplace—people expect to take the network with them
2003	Intelligent network agents ('knowbots') accepted as main mechanism for service provision
2007	Multimedia PDAs with speech interface become primary information access device
2009	Universal personal network service available
2010	Electronic newspapers, customised by subscribers, outnumber those printed on paper

nothing but information. The physical trappings used to deliver the information are, even after years of development, expensive and cumbersome. In the Information Age, that same information can be delivered further, faster and with a level of customisation to suit any reader. It is certainly possible. This book has covered some of the technology that would allow it to happen. But will it go this way? If so, when? Perhaps that is where the sort of planning outlined in this chapter will play a part, balancing the possible with the desirable.

Returning to the quotation that opened the book, the information superhighways will indeed have few speed limits and there is no turning back. There may be risk ahead but standing still is a sure recipe for oblivion.

However, the speed at which the highways are built and the direction they go in are yours to determine. We hope that this book gets you where you want to go . . . and that you miss the traffic jams on the way.

REFERENCES

Galliers, R. D. (1987) Information systems and technology planning within a competitive strategy framework. Pergamon Infotech Report.

Ives, B. and Learmouth, G. P. (1984) The information system as a competitive weapon. *Communications of the ACM*, December.

McFarlan, F. W. (1984) Information technology changes the way you compete. *Harvard Business Review*, May–June.

Norris, M., Rigby, P. and Payne, M. (1993) *The Healthy Software Project*. John Wiley & Sons.

Porter, M. E. and Millst, V. E. (1985) How information gives you competitive advantage. *Harvard Business Review*, July–Aug.

Rankl, W. and Effing, W. (1997) *Smartcard Handbook*. John Wiley & Sons.

Vitale, M. R. (1986) The growing risks of information systems success. *MIS Quarterly*, December.

Ward, J., Griffiths, P. and Whitmore, P. (1993) *Strategic Planning for Information Systems*. John Wiley & Sons.

Appendix 1:
Data Communications— Some Basic Concepts and Standards

Today we have naming of parts. Yesterday,
We had daily cleaning. And tomorrow morning,
We shall have what to do after firing. But today,
Today we have naming of parts . . .

Henry Reed (1946)

Data communications really began with the introduction of computers. During the early years, the 1960s and 1970s, the driver was the high cost of computer equipment. Time-sharing was the only way that most people could gain access to computers and time-sharing bureaux based on large mainframes were common with the telephone network used to provide switched access from remote 'dumb' teletype terminals.

However, advances in computers were rapid and hardware costs fell dramatically, mainly because of the development of integrated circuits. This exposed the shortcomings of the telephone network for data communications (it was very slow, error-prone, and made inefficient use of transmission) and stimulated the development of dedicated switched data networks. The fall in hardware costs and the introduction and growth of personal computers means that data communications are now motivated more by the desire to share information than to share equipment.

Work on switched data networks during the early 1970s led to the development of packet switching for data and the publication by CCITT in 1976 of X.25, the international standard for packet data services, in turn led to the widespread installation of packet switched networks, both private and public.

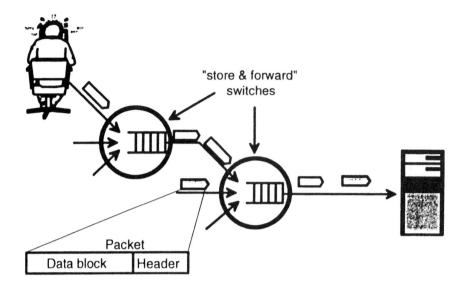

Figure A1.1   Packet switching

A1.1   PACKET SWITCHING

The distinguishing feature of circuit switching is that an end-to-end connection is set up between the communicating parties, and is maintained until the communication is complete. The most familiar example of a circuit-switched network is the telephone network; but for data communications packet switching is now firmly established as the favourite approach.

Communication between computers, or between computers and terminals, always involves the transfer of data in blocks rather than continuous data streams. As Figure A1.1 shows, packet switching exploits the fact that data blocks can be transferred between terminals without setting up an end-to-end connection through the network. Instead they are transmitted on a link-by-link basis, being stored temporarily at each switch *en route* where they queue for transmission on an appropriate outgoing link. Routing decisions are based on addressing information contained in a 'header' appended to the front of each data block. The term 'packet' refers to the header plus data block.

In fact the idea of store-and-forward switching in this way is older than packet switching. It has been used in a variety of forms to provide message switched services in which users could send messages (often very long messages) with the advantages of delayed delivery and broadcast options and retransmission if the message was garbled or lost in transmission. The distinctive feature of packet switching is that the packets, and consequently the queueing delays in the network, are short enough to allow interactive transactions.

The store-and-forward nature of packet switching brings a number of important advantages compared with circuit switching.

Because transmission capacity is dynamically allocated to a user only when he has data to send, very efficient use can be made of transmission plant. On the circuits between packet switches, packets from different users are interleaved to achieve high utilisation. This feature provided a strong incentive to introduce packet-switching where transmission costs are high, e.g. in North America where circuits tend to be long. Similarly, on his access circuit to the local serving switch the user can interleave packets for different transactions, enabling communication with numerous remote terminals simultaneously.

Because the network effectively buffers communicating terminals from each other they can have access circuits of different speeds. However, to prevent the faster terminal from overloading the slower, end-to-end 'flow control' needs to be imposed to limit the rate at which packets may be sent. This is generally based on permission-to-send being returned to the sending terminal by the receiving terminal, independently for each direction of transmission (see the description of X.25 below for an example of how this flow control is achieved in practice).

Very low effective error rates can be achieved. By adding an error-detecting checksum (in effect a complex parity check) to the end of each packet errors can be detected on a link-by-link basis and corrected by retransmission. At each switching node *en route* packets would be stored until their correct receipt is acknowledged. This arrangement can also provide a high degree of protection against failures in the network since packets can be rerouted to bypass faulty switches or links without the user being aware of a problem.

Network flow control and congestion control

In a packet-switched network packets compete dynamically for the network's resources (buffer storage, processing power, transmission capacity). A switch accepts a packet from a terminal largely in ignorance of what resources the network will have available to handle it. There is always the possibility therefore that a network will admit more traffic than it can actually carry, with a corresponding degradation in service. Controls are therefore needed to ensure that such congestion does not arise too often and that the network recovers gracefully when it does.

The importance of effective flow and congestion control is illustrated in Figure A1.2. For a given delay performance, a packet-switched network can be regarded as having a notional maximum throughput, Max packets per second. An ideal network would accept all offered traffic up to its throughput limit: beyond this some traffic would be rejected but the network's throughput would stay high. In a network with poor flow and congestion control, however, the throughput, although initially increasing with offered traffic, would eventually fall rapidly as a result of congestion. In practice good flow control maintains high throughput but leaves an adequate margin of safety against network failures and the imperfect operation of practical algorithms.

Any flow control imposed by the network is in addition to end-to-end flow control operated by the terminals to match differences in speed. In fact

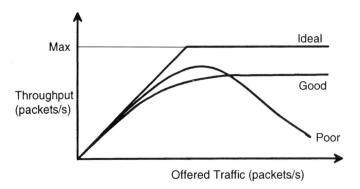

Figure A1.2 The importance of network flow control

end-to-end flow control is itself effective in avoiding network congestion, since the increase in cross-network delay caused by the onset of congestion will delay the return to the sending terminal of permission-to-send, effectively reducing the rate at which it may offer further packets to the network.

Connectionless and connection-oriented services

The simplest form of packet-switched service is the connectionless or datagram mode in which each packet, or datagram, is regarded as a complete transaction in itself. There is no sense of a connection being set up before communication begins and the network treats each packet independently: there is no sense of packet sequence. The connectionless mode is generally used in local area networks. SMDS, designed primarily to provide LAN-interconnection over long distances, is also a connectionless service.

Many applications, however, involve the transfer of a sequence of packets, for which a connection-oriented approach is more appropriate in which a connection is established by an initial exchange of signalling packets between the communicating terminals. Data transfer then takes place and at the end of the session the connection is cleared. During the data transfer, or conversation, phase the network tries to create the illusion of a real end-to-end connection using store-and-forward switching. But to distinguish them from 'real' connections, they are referred to as virtual connections or circuits (it should be noted that circuit-mode services are intrinsically connection-oriented).

Packet switching and circuit switching compared

A circuit-switched network such as the PSTN provides an end-to-end connection on demand, provided of course that the necessary network resources are available. Once established, the users have exclusive use of the connection for the duration of the call. The connection's end-to-end delay is small (unless satellite circuits are used) and constant, and other users cannot interfere with the quality of communication. But if the required network resources are not available at call set-up time, the call is refused—in effect the call is lost. Circuit-switched networks are, in the jargon, 'loss' systems.

In contrast, in a packet-switched network packets queue for transmission at each switch. The cross-network delay is therefore variable, depending as it does on the volume of other traffic encountered *en route*. Packet-switched networks are, in the jargon, 'delay' systems. In effect a packet-switched network is a queue of queues, and queueing theory is a basic part of the network designer's toolkit. In practice the queueing problems presented by real packet networks cannot be solved analytically, and performance modelling generally involves a combination of analytical and simulation methods.

Table A1.1 contrasts the main characteristics of circuit and packet-mode services.

Table A1.1

	Circuit-switched	Packet-switched
Delay	short and fixed	longer and variable
Error rate	same as connection	very low
Simultaneous calls	no	yes, packet interleaved
Speed matching	no	intrinsic
Transmission efficiency	low to medium	medium to high
Resilient to n/w failures	no, calls are lost	yes, the data is rerouted

A1.2 THE OPEN SYSTEMS INTERCONNECTION (OSI) REFERENCE MODEL

Unlike telecommunications, the IT industry has never been subject to stringent regulation or monopoly supply. Consequently, as data communications developed the trend was towards diversity rather than standardisation driven by the desire for product differentiation. Every IT supplier developed his own set of communications protocols, or their own variations of someone else's, and interworking equipment from different suppliers was usually difficult and often impossible. The reference model for data communication arose from the need for a clear framework within which to develop data communication standards as a basis for widespread inter-operability—that is, open systems interconnection (OSI).

The layered approach arose naturally as a structure for 'dividing and conquering' the growing complexity. Based on ten guiding principles, a seven-layer reference model was developed by ISO (ISO 7498), and now also adopted by ITU-TS (X.200), as shown in Figure A1.3. The idea is that each layer in the protocol stack provides a defined service to the layer above it. This definition is in implementation-independent terms so that the technology at each layer can, in principle, evolve independently of that at other layers.

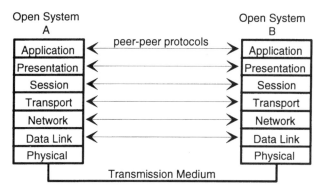

Figure A1.3 The OSI seven-layer reference model

Starting at the bottom is Layer 1, the physical layer, which is concerned with transmission. Specifically it provides 'mechanical, electrical, functional and procedural means to activate, maintain, and deactivate physical connections for bit transmission' between data link layer entities. It embraces all types of transmission media: copper wires, co-axial cable, optical fibre, radio, etc.

Layer 2, the data link layer, is responsible for getting data units from Layer 3, the network layer, to the remote peer network layer process. It is concerned with getting the physical layer to set up and control the necessary physical connections (more than one physical connection may be needed between peer data link layer processes to achieve the desired throughput or reliability. The data link layer would spread the data units over the physical connections and collect them together again at the far end). It implements: sequence control to ensure that network layer data units are received in the same order in which they are transmitted; flow control to ensure that the sending end does not transmit data units faster than the receiving end can deal with them; and error detection and error recovery. It may also include a multiplexing/demultiplexing function to enable a single data link layer to carry packets belonging to more than one network layer connection.

We will see in section A1.4 on LANs that, based on the same reasoning, it has also been useful to define distinct sublayers within the data link layer.

Layer 3, the network layer, provides two types of network service, the connection-mode network service (CONS) and the connectionless-mode network service (CLNS). For the connection-mode service the network layer provides 'the means to establish, maintain and terminate network connections between open systems . . . and the functional and procedural means to exchange network service data units [packets] between transport entities over network connections'. In other words the network layer sets up and manages the network connection between the communicating open systems. This connection may pass through more than one network, and will involve network layer protocols appropriate to the networks used, which may be packet-switched (including X.25, LAN or connectionless network), or circuit-switched (including the telephone network, ISDN and X.21).

The purpose of Layer 4, the transport layer, is to make sure that the data units from the session layer are carried across the intervening network(s) without errors, without loss, without duplication and in the right order, while concealing network-dependent aspects from the higher layers. It uses the services provided by the network layer which is network-dependent. The transport protocol used in a given situation will therefore depend on the nature of the network(s) used. If the error rate demanded by the session layer is better than the underlying network can provide, then the transport layer will operate an error recovery protocol. But if the underlying network can provide robust enough data transfer (such as X.25 described below) then there is no need for the transport layer to perform error correction as well.

Layer 5, the session layer, shows no network-dependent aspects, and has the general role of ordering and synchronising communications over the reliable connections managed by the session layer. The session layer provides the capability to mark points in the dialogue between the higher layers so that synchronisation can be managed and re-established if necessary, for example following a failure of one of the end systems.

Layer 6, the presentation layer, is concerned with the representation of the data to be transferred between the application layer processes. This may embrace aspects of encryption, text compression, or conversion between the syntax and/or data formats used by the two end systems (such as conversion between EBCDIC and ASCII), which may involve negotiation with the remote peer presentation layer to establish the appropriate syntax. It is not concerned with what the data means, merely its representation.

Layer 7, the application layer, provides the end system applications with the 'window' into the open systems environment. The OSI reference model does not include standardisation of the applications themselves, but on how the applications communicate with each other. For example, an application may be concerned with CAD/CAM (computer-aided design/manufacture) and a typical interaction of remote CAD/CAM end systems would involve file transfers. The open systems reference model would be concerned with the file transfers but would have no knowledge or visibility of the CAD/CAM application itself.

A1.3   X.25: THE INTERNATIONAL STANDARD FOR PACKET DATA SERVICES

CCITT published X.25, the international standard for packet switched data services, in 1976 and most countries now have a public network infrastructure that provides X.25 services. The complete standard runs to over 150 pages (and refers to many other standards) and is not light bedtime reading! The following brief description can therefore provide only a sketchy outline. It is included because it is in widespread use and because it illustrates in a specific way some of the features introduced in section A1.1, such as end-to-end flow control.

The original 1976 version covered only virtual circuit services, but continued arguments for a simple datagram option led to the inclusion of a datagram service in the 1980 revision. But nobody seemed interested and it was taken out again in the 1984 revision (CCITT standards tend to be reviewed every four years). The 1984 revision also took account of the OSI reference model and provisions were included for X.25 to support the OSI connection-oriented network service (CONS).

One of the most striking features of X.25 is the way that it is structured into three distinct layers, as shown in Figure A1.4. Though familiar now because of the OSI reference model, in 1976 this was a significant innovation. The three layers correspond to the bottom three layers of the OSI model: the physical layer; the link layer (equivalent to the data link layer), and the packet layer (equivalent to the network layer). X.25 actually defines the service at the interface between the terminal and the network and the physical layer is concerned with the circuit connecting the terminal to the serving packet switch; the data link layer provides for the error-free transfer of packets over this access circuit; and the packet layer defines a virtual circuit protocol capable of supporting multiple simultaneous virtual calls.

Being a virtual circuit protocol, there is a call set-up phase, a data transfer phase, and a call clear phase. The user can have permanent virtual circuits (PVCs), which are set up by the network operator on a subscription basis and can be considered to be permanently in the data transfer phase, or switched virtual calls (SVCs), which are set up and cleared as required by the user by an initial exchange of signalling packets, as shown in Figure A1.5. A user may have a (potentially large) number of virtual circuits set up simultaneously to different destinations: these may be any combination of PVCs and SVCs.

X.25 Call set-up and clear

The calling terminal forms a CALL REQUEST packet and sends it into the network. This call request packet contains the full address of the called terminal (we will see later what these addresses look like) and is routed by the network to the appropriate terminal, where it is delivered as an INCOMING CALL packet. The called terminal accepts the call by returning a CALL ACCEPTED packet. The network routes this back to the calling terminal where it is delivered as a CALL CONNECTED packet. At this point the virtual call enters the data transfer phase.

At virtual call set-up time an identifier is associated with the call for its duration. This is known as the logical channel number, and it is used in all packets belonging to the call to distinguish them from packets belonging to other calls which the terminal may have in progress simultaneously. During the data transfer phase, packets do not carry the full address of the remote terminal, they just contain the logical channel number.

At any time during the data transfer phase either terminal can clear the call. Figure A1.5 shows the call being cleared by the calling terminal. It sends a CLEAR REQUEST packet into the network. This is delivered to the remote terminal as a CLEAR INDICATION packet. The appropriate response to this by the called terminal is a CLEAR CONFIRMATION packet which the

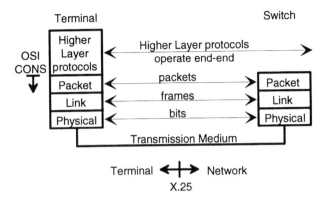

Figure A1.4   The X.25 protocol stack

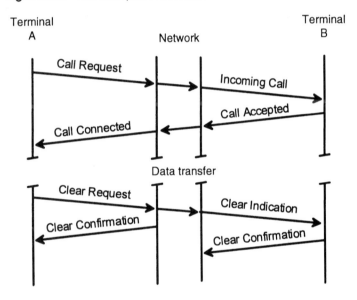

Figure A1.5   X.25 virtual circuit set-up and clear

network delivers to the calling terminal, again as a CLEAR CONFIRMATION packet.

X.25 packet layer data transfer

User information is carried in DATA packets with the format shown in Figure A1.6. One of the important features of a virtual call is that it preserves packet order: packets are delivered in the same order in which they are sent. To achieve this, each DATA packet carries a send sequence number P(S).

A flow control procedure operates during data transfer to control the rate at which packets may be sent. This flow control operates independently for each direction of transmission and uses what is known as a sliding window

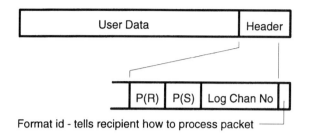

Figure A1.6   X.25 data packet format

mechanism. For each virtual call a terminal may send up to an agreed maximum number of packets into the network, after which it must wait for permission-to-send before it may retransmit more.

This permission is returned in the form of acknowledgements to packets sent previously. For this purpose a receive sequence number P(R) is returned, which is set equal to the send sequence number expected in the next packet to be received, i.e. one more than the send sequence number of the last packet that has been acknowledged.

The window refers to the maximum number of packets that may be sent, awaiting acknowledgement. As packets are transmitted the window closes. As they are received and acknowledged it is opened. Figure A1.7 shows an example of how the send and receive sequence numbers control the flow of packets. It should be noted that packets are not necessarily acknowledged one at a time. One receive sequence number may be used to acknowledge receipt of a group of packets.

As part of the flow control procedure short (3-octet) packets may be used in addition to DATA packets. They are RECEIVE NOT READY (RNR) packets which are sent to indicate a temporary inability to accept further DATA packets on the call, and RECEIVE READY (RR) packets which are sent to indicate that the terminal is ready to accept further DATA packets on the call. If a terminal receives an RNR packet it should immediately stop sending DATA packets on this call until the temporary busy condition is cancelled, which may be done by sending an RR packet.

Both RNR and RR packets contain receive sequence numbers and play an important part in the sliding window flow control mechanism (in addition to their more direct role of stopping and starting packet flow). In particular, if a terminal (or a switch) wishes to acknowledge receipt of DATA packets but has no DATA packet ready to send, it will generally use a RECEIVE READY packet to carry the P(R) needed to perform the acknowledge.

This is illustrated in Figure A1.7, which gives a simple illustration of the procedure. We assume that the call has just entered the data transfer phase and that the maximum window size is 2.

The first data packet carries a send sequence number of 0 and a receive sequence number of 0. There are no data packets from the remote terminal waiting to be delivered on which the acknowledgements could be 'piggybacked', so the network sends a RECEIVE READY packet to acknowledge

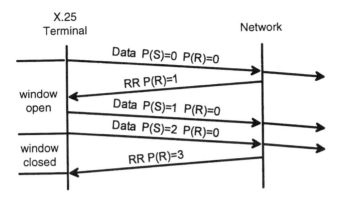

Figure A1.7   X.25 packet layer data transfer

receipt of the first data packet. This RR packet carries a receive sequence number of 1 to indicate that it is acknowledging the data packet with send sequence number 0.

The terminal then sends out two DATA packets in quick succession, with send sequence numbers 1 and 2 (the receive sequence numbers are still 0 because we have not received any DATA packets from the remote user that need acknowledging). The window then closes and the terminal has to stop sending DATA packets until at least one of the outstanding DATA packets has been acknowledged. In the example this is again done using an RR packet generated by the network, and it carries receive sequence number 3 to indicate that it is in fact acknowledging both DATA packets 1 and 2. At this point the window opens and the terminal can send more DATA packets.

That is how normal packet layer data transfer takes place in X.25. But the real world does not always work smoothly and in practice things sometimes go wrong. For example a packet might get lost in the network. This would cause a packet to be received out of sequence; that is, it would have a send sequence number different from that expected. X.25 includes procedures to recover from such errors.

There is a REJECT procedure in which a REJECT packet is used to request the retransmission of DATA packets beginning from the indicated receive sequence number. There is a RESET procedure used to recover from more serious error situations. This resets all sequence numbers to 0 and resumes data transfer: packets may be lost during RESET. Finally, there is the somewhat more drastic RESTART procedure which is applied to all virtual calls that a terminal has in progress. It simultaneously clears all SVCs and resets all PVCs, in effect re-initialising the complete packet layer process.

Also, in the real world it is sometimes desirable for a terminal to be able to send information to the remote terminal without observing the flow control procedure. This can be done in X.25 using the interrupt feature whereby the terminal sends an INTERRUPT packet. Only one INTERRUPT packet can be outstanding (that is unacknowledged) at a time, and the interrupt procedure has no effect on normal data transfer. Up to 32 octets of data can be conveyed in an INTERRUPT packet.

X.25 data link layer

The X.25 packet layer depends on a reliable means of transferring packets between the terminal and the network. As shown in Figure A1.4, the X.25 data link layer provides this.

The data link layer protocol includes procedures for setting the link up, for data transfer, and for clearing the link down. During data transfer the data link layer protocol operates sequence control and flow control using the same type of sliding window flow control procedure as outlined above for packet layer data transfer, but applied across all virtual circuits in progress, not on a per virtual circuit basis. Only one pair of sequence numbers is used by the link layer to cover all packet traffic.

But the data link layer also operates error control to overcome the effects of errors that may been introduced in transmission. Each X.25 packet is carried by the data link layer as the information field of a link layer frame. Each frame has a trailer containing a cyclic redundancy checksum (CRC), in effect a very comprehensive form of parity check. At the receiving end of the link a new CRC is calculated for each frame. This fresh CRC is compared with that contained in the trailer of the received frame. If they correspond then it may be presumed that there are no transmission errors. If they do not correspond the error is corrected by retransmission.

X.25 physical layer

This covers the mechanical and electrical aspects of the interface and the procedures needed to set up and maintain the physical link between the terminal and the network. X.25 explicitly includes X.21, X.21bis, V.24 and X.31 (which covers access via an ISDN).

X.121 numbering

Terminal addresses in X.25 signalling packets are based on X.121, the international numbering plan for public data networks, and have the format shown in Figure A1.8.

Each country has a three-digit data country code (DCC), administered by ITU. The fourth digit, the network digit, identifies specific data networks within a country. Together the DCC and network digit form the data network identification code, or DNIC, which uniquely identifies a network. If a country has more than ten networks it is necessary to assign more than one country code; the UK for example has been allocated four data country codes, 234 to 237, and the USA has seven, 310 to 316.

In addition to the DNIC there is a network terminal number (NTN) of up to 10 digits. If there is only one data network in a country the network digit is often combined with the NTN to form what is known as the national number.

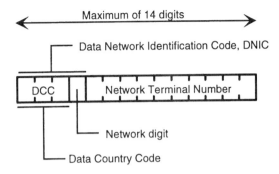

Figure A1.8 The X.121 numbering scheme

A1.4 LOCAL AREA NETWORKS (LANs)

In addition to the development of wide area packet switched data networks, the late 1970s also saw the development of local area networks. Again these were originally driven mainly by the desire to share equipment, either by providing access to a shared mainframe or mini-computer or by sharing peripherals such as printers or servers. But the same forces shaping wide area data communications have also been at work in the local area and LANs are now used to share information rather than equipment.

By 1980 the LAN industry was clearly established as a growing market and in that year the IEEE 802 Committee began its work on developing LAN standards. Three main types of LAN have emerged as IEEE standards, the so-called CSMA/CD LAN (carrier sense multiple access with collision detection, based closely on the Xerox Corporation's Ethernet), the Token Ring and the Token Bus. The IEEE also developed a framework for the definition of LAN standards that has contributed greatly to their success in promoting interoperability. In particular they define two distinct sublayers within the data link layer, as shown in Figure A1.9: the logical link control sublayer (LLC) and the medium access control sublayer (MAC).

The MAC sublayer is concerned with how a station gains access to the physical medium in order to send packets; it is different for each of the types of LAN. Three IEEE standards 802.3, 802.4 and 802.5, respectively, define the MAC sublayer (together with the physical layer) for each of the LAN types as shown.

The LLC sublayer, defined by 802.2, is concerned with providing the network layer with the data link layer service over the MAC sublayer; that is, setting up logical links between stations and, when appropriate, operating flow control and error recovery. Together the LLC and MAC sublayers provide the OSI data link layer service.

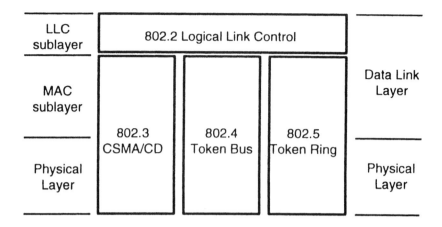

Figure A1.9   IEEE 802 LAN standards

The CSMA/CD (Ethernet) MAC sublayer: 802.3

The 802.3 LAN is physically organised as a bus, appropriately terminated to control reflections, to which the stations are attached. The basic operation of the MAC protocol is that when a station has a packet (more correctly, data unit) to send it first 'listens' on the bus, and if no other transmissions are in progress it sends the data unit. If another station is already transmitting the station waits until the bus has become free before trying again.

When a station has started to transmit a data unit it continues to listen for a while to check that another station has not started transmitting at the same time, or nearly the same time, that is, collided. If what a station hears is not the same as what it is transmitting, a collision is indicated. In this way all colliding stations will back off and wait before trying again. To minimise further collisions each station will wait for either a random period or a period preset for the station.

A range of implementations have been defined for the physical sublayer offering a wide range of price and performance. These include coaxial cable giving a data rate of 10 Mbit/s, and twisted-pair cable giving either 1 or 10 Mbit/s data rate. The CSMA/CD LAN is best suited to comparatively light and bursty loading. If loaded too heavily, collisions become more frequent and stations spend more of their time backed off waiting to transmit.

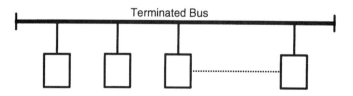

Figure A1.10   CSMA/CD (Ethernet) LAN

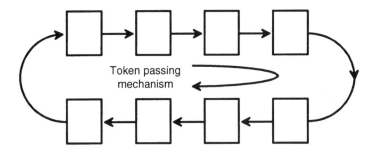

Figure A1.11   The Token Ring LAN

The Token Ring MAC sublayer: 802.5

The 802.5 LAN avoids collisions by using a token to synchronise access to the shared transmission medium, that is, the ring. The token is a unique bit pattern that is transmitted around the ring. In order to gain access to the ring to send a data unit a station must wait until it receives the token. Only then may it transmit its information. The data unit will pass through all stations on the ring, and only the station to which a data unit is addressed should retain a copy. Having travelled completely around the ring the data unit will, errors permitting, arrive back at the originating station (it will also include an indication from the addressed station that the data has actually been copied) which removes the data unit and transmits the token to the next station. In this way each has station gets its opportunity to send.

The Token Ring LAN is better than the CSMA/CD LAN for a heavily loaded network. The absence of collisions (under normal conditions) permits high utilisation of the ring and limits the time any station has to wait to transmit. Though not described here, there is also a priority scheme that permits different levels of priority to be associated with different data units. Through the support of IBM the Token Ring is now becoming very popular. It offers data rates up to 16 Mbit/s.

The Token Bus MAC sublayer: 802.4

The 802.4 LAN uses a similar principle of token passing as the Token Ring just described above. But since it is based on a bus rather than a ring, the token has to be explicitly passed from one station to the next as an addressed transaction. The stations on the bus are actually arranged as a logical ring; in addition to its own address, each station on the bus knows the address of its logical predecessor and its logical successor. So it receives the token from its logical predecessor and, after transmitting any data units it may have to send (up to a time-out limit), it passes the token on to its logical successor.

This has similar features to the Token Ring LAN, but offers more flexibility

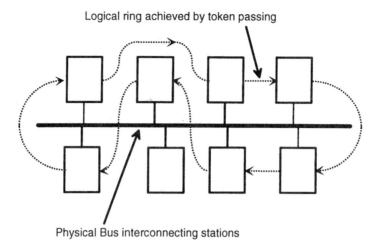

Figure A1.12   The Token Bus LAN

in that the binding between physical location and position on the logical ring can be changed without physical intervention by centralised management. However, there is a distinct overhead in passing the token and maintaining the integrity of the logical ring. Data rates up to 10 Mbit/s are specified for coaxial cable and up to 20 Mbit/s for optical fibre. Under the name manufacturing automation protocol (MAP) the Token Bus is used as a flexible and robust factory networking standard by manufacturing companies.

The LLC sublayer: 802.2

The LLC sublayer operates a protocol with its remote peer which in its full glory supports three distinct services which the network layer may select as required. Type 1 is an unacknowledged connectionless service: it includes neither flow control nor error recovery. Type 2 is a connection-oriented service for which a data link connection is set up between the communicating stations: it provides for sequence control, flow control using a sliding window mechanism and error recovery. Type 3 is an acknowledged connectionless service which provides an acknowledgement to each data unit transmitted.

A LAN station is classed as Class I if it supports only the Type 1 service, as Class II if it supports Types 1 and 2, as Class III if it supports Types 1 and 3, and as Class IV if it supports all three Types. All LAN stations operate the Type 1 service, which is very effective in the comparatively error-free LAN environment despite its paucity of features, or perhaps because of its paucity of features! The corresponding absence of overheads enables high throughput to be achieved.

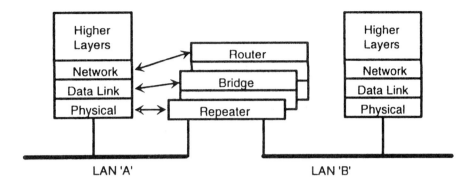

Figure A1.13   Repeaters, bridges and routers

A1.5   LAN INTERCONNECTION

As the place of LANs became established as part of a company's enterprise network infrastructure, the need developed to extend them and interconnect them with other LANs located both locally and remotely. A number of devices have been developed to achieve this, as shown in Figure A1.13.

Repeaters

The simplest such device is the repeater, which operates at the physical layer in OSI terms. It is used to extend the physical reach of a LAN simply by regenerating the electrical (or optical) signal, overcoming any attentuation or dispersion. But it should be recognised that distortions of this sort are not the only factor that limits the physical reach of a LAN: transmission delay also has an impact, and repeaters can only add delay, they cannot remove it. So there is a practical limit to the number of repeaters that may be used in a network and in the degree of network extension that can be achieved.

Because the repeater operates at the physical layer it can only be used to interconnect LAN segments if they are of the same type; that is, CSMA/CD bus to CSMA/CD bus, or token ring to token ring.

Bridges

Bridges operate at the data link layer and are therefore transparent to higher-layer protocols. They are available in a wide range of functionalities, but come in two general flavours: transparent and translating. A transparent bridge is used to interconnect LANs that have similar data link layers; CSMA/CD to CSMA/CD, for example. The more recently developed translating bridges, on the other hand, can interconnect LANs with different MAC sublayers, such as CSMA/CD to token ring, and perform the necessary translation between the MAC sublayer frame formats. In effect bridges preserve the illusion that interconnected LANs are simply one large LAN.

In addition to any translation that may be needed the basic functionality of the bridge is one of filtering and forwarding. By observing the source and destination addresses of passing data units the bridge maintains address tables so that it knows which stations are local (that is, connected to the same LAN or LAN segment) and which ones are on other LANs or segments. Local traffic is kept local, it is not passed to adjacent segments. In this way the bridge isolates a LAN segment from unnecessary traffic, which increases the LAN capacity available for the local stations.

Some bridges are designed for local use; that is, for interconnecting LANs that are local to each other. But others are designed to interconnect remote LANs via a wide area network, as shown in Figure A1.14. The difference is that the remote bridge has to arrange for the LAN data units to be carried by whatever wide area protocol is operated by the intervening WAN. This may be a simple leased line connection. It may be an X.25 virtual circuit. Or it may be Frame Relay, SMDS or, in the near future, ATM.

The normal approach is to encapsulate the LAN data unit in the data unit used by the wide area network; that is, the complete LAN data unit is inserted into the user information field of the WAN data unit at the sending end and extracted at the receiving end. But since a bridge may carry a number of network layer protocols over the same WAN connection, it is generally necessary to include an indication of which network layer protocol is being carried in a particular data unit so that the encapsulated LAN data unit can be given the appropriate form of processing at the far end. How this is achieved for a Frame Relay WAN is described in Chapter 3.

Routers

Routers operate at the network layer and are therefore sensitive to the network layer LAN protocols actually used (such as Xerox's XNS, Digital's DECnet, Novell's IPX, AppleTalk, TCP/IP, and so on). Some routers are single-protocol devices that will only communicate sensibly with other routers using a specific LAN protocol. Others are more flexible (and more complex!) and can handle multiple LAN protocols.

Like bridges, routers also maintain address tables and perform filtering. But they route on the basis of network layer addresses, not the MAC sublayer addresses used by bridges, and communicate interactively with other routers to exchange routing information; they can support complex meshed networks of interconnected LANs in a way that bridges cannot. In general routers provide for interconnection via wide area connections.

Gateways

The term gateway is often used generically to refer to any type of interworking

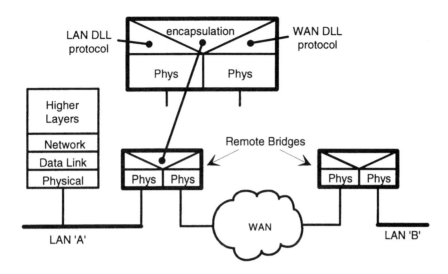

Figure A1.14   Remote bridges

device such as a bridge or router. But, it properly refers to an interworking device that operates at a higher layer than the network layer.

REFERENCES

General

Davidson, P. R. and Muller, N. J. (1992) *Interworking LANs.* Artech House.
Griffiths, J. M. (1992) *ISDN Explained.* John Wiley & Sons.
Knowles, T., Larmouth, J. and Knightson, K. G. (1987) *Standards for Open Systems Interconnection.* BSP.

Standards

ISO/IEC

ISO 7498 OSI Basic Reference Model (there are numerous additional parts to this, covering such aspects as naming, addressing and management). There are also numerous other standards which develop the reference model further, such as ISO 8326, and they are not all listed here.
ISO 8326 Basic Connection-Oriented Session Service Definition
ISO 8072 Transport Service Definition
ISO 8348 Network Service Definition
ISO 8473 Protocol for Providing the Connectionless-mode Network Service

ISO 8878 Use of X.25 to Provide the Connection-Mode Network Service
ISO 8886 Data Link Service Definition

ANSI/IEEE

802.2 Local Area Networks—Logical Link Control (also published as ISO 8802-2)
802.3 Local Area Networks—CSMA/CD access method and physical layer specifications (also published as ISO 8802-3)
802.4 Local Area Networks—Token-passing bus access method and physical layer specifications (also published as ISO 8802-4)
802.5 Local Area Networks—Token Ring access method and physical layer specifications (also published as ISO 8802-5)

CCITT/ITU-TS

V.24   List of definitions for interchange circuits between data terminal equipment (DTE) and data circuit-terminating equipment (DCE)
X.21   Interface between DTE and DCE for synchronous operation on public data networks
X.25   Interface between DTE and DCE for terminals operating in the packet mode and connected to public data networks by dedicate circuit
X.31   Support of packet mode terminal equipment by an ISDN
X.121 International numbering plan for public data networks
X.200 Reference model of open systems interconnection for CCITT applications (X.211 to 216, respectively, give service definitions for the physical layer service, data link layer service, up to the presentation layer service)

Appendix 2
Distributed Computing— Some Basic Concepts and Standards

There are some people who believe that they can get by without even the basic technical facts.
In much the same way, a donkey can get by without legs—it simply hauls itself forward with its teeth

Mike Shields

Human organisations are distributed by their very nature. People work in different places and information is acquired and stored at different locations. It is often convenient and sometimes necessary to locate computers close to the information they process for reasons of performance, accessibility or security. The growth in local area networks which give fast controlled access to a range of local resources is a testament to this. It has been argued in the main text that people increasingly rely on computers in their work and require good response and availability from the computers they use. In time, the nature of a computer system comes to reflect the human organisation it serves. The move towards more distributed, information-intensive working needs to be reflected in powerful distributed computing.

This appendix outlines both the technical and organisational challenges of distributed working. Some guidelines for organising effective distributed teams are given at the end of the appendix. First, we take a look at the prospects for flexible and reliable distributed computing, the basis for virtual organisations.

Given that human organisations are not static but are subject to change, this must be accommodated by the computers used to support the organisation.

As a consequence, systems will have to be provided as a collection of interchangeable components, rather than as a monolithic single resource bound to one supplier or configuration.

As organisations seek to add increasingly complex functions to their systems, suppliers will find it increasingly difficult to manufacture, sell and support a full range of system components. It is reasonable to assume that suppliers will cater for specific needs and will, therefore, be obliged to co-operate with other suppliers (or their products) when they are called on to assemble a working system.

Large organisations engage in many diverse activities, each of which can be supported by a computer system. For example, a manufacturing company may have computerised payroll, order handling, design and sales support systems. To make best use of these systems they need to be linked to provide an enterprise resource. This is fairly straightforward when all the systems are of the same type, but the diversity of requirements usually demands a diversity of hardware and software support. Distributed processing provides a route to achieving the integration required by the user without restricting the freedom to choose the most appropriate technology for each system component.

In a nutshell, these are the reasons why the computing world has had to meet the telecommunications world. It is a convergence driven by need not choice. That need (as we have already said) is in the simplest terms the user wanting to have the facilities now commonly available on an LAN across all of the network.

The two approaches to achieving this are distinct and are largely isolated. Those who are familiar with both sides are few and far between (but precious). In the world of telecommunications the main drive is to make it easier for the user to transfer more and more data from one place to another. In the computing world, this same end has been tackled through 'distributed processing'.

Since the language, culture and background in computing and telecommunications circles are somewhat different, the same issue has been tackled in (at least) two rather different ways. Even within the computing community, there are several different ways of getting to the same goal. The main factions to have emerged over the last ten years are as follows.

- The communications community, which focuses on the OSI model as a universal basis for two computers to communicate in a defined fashion (e.g. file transfer, message transfer, virtual terminal, etc.)

- The distributed operating system community whose aim has been to host the functions of a particular operating system on a network of processors.

- The RPC community, which focuses on general applications within the paradigm of procedure calls. This body is influential in that it addresses the full range of distributed processing issues.

- The network applications community, which is concerned with things you can do over a network. This embraces Internet, whose networked news (Usenet) has been in use since 1979, with subsequent applications (gopher, WAIS, World Wide Web) following.

The different directions taken by these various communities has done little to achieve a consistent approach to distributed computing; the RPC world looks on OSI standards as too heavy and specialised, the operating system people are seen as low-level and as not addressing heterogeneity. Meanwhile, the vendors have gone their own way by building distributed versions of their products, not necessarily interchangeable with other products; DBMS vendors provide distributed SQL, TP vendors distribute transactions, etc.

Despite these variations in opinion and practice, there is some common ground. We now look at some of the basic concepts and standards that have evolved over the last few years in distributed computing.

This is not intended to be anything like a detailed review—that would require several more books. It would also be somewhat transitory information, as the computing industry tends overall to progress somewhat more rapidly than the traditional telecommunications market. This year's fad may become next year's millstone; the watchwords for the rest of this chapter are *caveat emptor*: buyer beware, volatile technologies under discussion!

There are some general principles, though, that do characterise what is happening in this area. Before moving on to the particular techniques that are now emerging, consider the enablers for distribution and the standards that embody them.

A2.1   DISTRIBUTED PROCESSING

Perhaps the most important point is that the whole thrust to date has been on making a set of computers co-operate. The connection between the computers in question has been, to some extent, assumed, and here we have an important differentiator between this and the previous appendix: the telecommunications focus has been on the transport of data; but the computing focus has been on co-operation between computers.

In practice, this means that the distributed processing work has focused more on logical connection than physical. In terms of the OSI 7 layer model introduced earlier, we are now concerned with the upper layers. Some of the protocols and standards introduced through this book are placed on the model in Figure A2.1.

Each of the layers in the diagram has a specific purpose, as follows:

- Physical layer, concerned with how to transmit the zeros and ones that constitute digital information from one place to another

- Link layer, concerned with getting those bits reliably from one node to the next. Issues such as framing, error detection and correction, and flow control are dealt with here

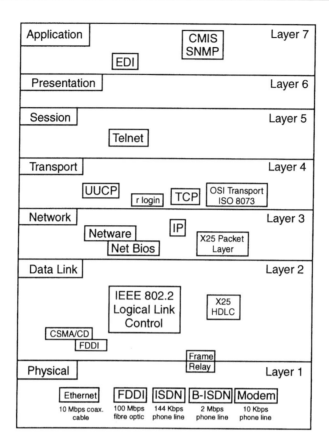

Figure A2.1   A lightly populated Open Systems Interconnection 7 layer model

- Network layer, concerned with providing an end-to-end path between client and server; this is where, for instance, global addressing and routing is dealt with

- Transport layer, which operates end-to-end and assumes that a reliable connection is established

- Session layer, which provides dialogue separation and control for those applications that need it; for client/server applications, this is an inherent part of the design

- Presentation layer, which provides a mechanism for negotiating forms of representation (known as transfer syntax) for a given message content

- Application layer, which holds the remaining application dependent functions required (e.g. CMIS as a general network management tool).

There is considerable evolution within the first three layers, those that define the network service. As new technologies come along, so the realisation within the framework changes. The next three layers are essentially complete

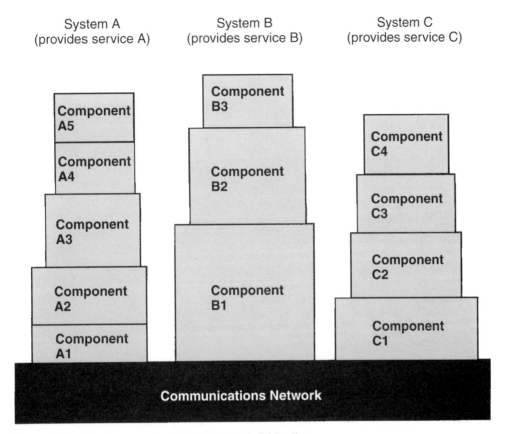

Figure A2.2 The 'connected stack' view of distribution

with only minor changes and additions to functionality expected. The top layer (application) is, by its nature, open ended. The likely development here is the provision of tools that can be used by a wide range of applications.

A very large proportion of the output in distributed processing has been focused on layers 4 and 5 of the model. The aim is to define how computers from different manufacturers can co-operate, how they identify each other, how they process information without a traditional central controller, etc. The overall thrust of open systems is to move from a situation (depicted in Figure A2.2) where you can connect systems together but where interoperability is (at least) awkward with information flowing down one stack to be sent to the next stack, where it has to percolate to the top to complete its journey.

The target is to ensure that communications and information handling, presentation, etc., is common across different system types. In effect, this means having peer-to-peer operation of the seven-layer model, rather than each stack behaving as a separate entity with the only route in being through the lowest level of connection. The attraction is greater flexibility and efficiency.

Despite this, there is no single agreed definition of the terms distributed processing or distributed system. At a very simplistic level, a system becomes distributed if the functions that are performed are not contained within a

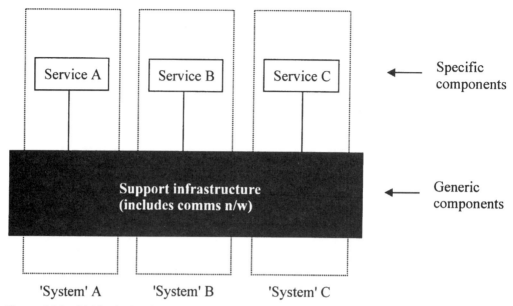

Figure A2.3 Distributed systems in co-operation

single processor. The benefits of this limited form of distribution are minimal. A more precise, but still high-level, definition is that a distributed system is that which is required to facilitate distributed processing. A working description of distributed processing that we would put forward is that 'Distributed processing comprises a collection of services and resources which are located on a variety of systems linked by one or more network to form a coherent whole'.

In other words, a distributed system comprises interworking applications supported by geographically separate computing platforms. So what does this mean, and how is a distributed system different from any other?

A number of simplifying assumptions commonly made when constructing centralised systems cannot be made when designing distributed systems. This changes the manner in which application design has to be approached. Among the most important assumptions which need to be carefully re-examined on moving from a central to a distributed systems are the following.

- Central control: a distributed system may be spread across a number of autonomous authorities, with no single authority in overall control. This may lead to problems of consistency between domains.

    central control → autonomy

- Single global name space: in distributed systems, which may arise from mergers of pre-existing systems, context-relative naming schemes are required in order to interpret names unambiguously across different administrative boundaries.

    global naming → federated naming

- Global shared memory: attempting to construct the illusion of shared memory spanning nodes of a large-scale distributed system is not practical.

  global shared memory → locally encapsulated state

- Global consistency: in distributed systems, consistency of state and data may converge more slowly than in non-distributed systems.

  global consistency → weak consistency

- Sequential execution: program execution on separate processors in a distributed system by default takes place concurrently.

  sequential execution → concurrent execution

- Failure: in the event of some part of a distributed system failing, the remainder can often continue operating without it. Indeed, in a large system it is unlikely that all parts will ever be operational simultaneously. A central system is either working or broken.

  vulnerability → fault tolerance

- Locality of interaction: in distributed systems, interactions may be either local or remote, with consequent implications for communication delay and reliability.

  local → remote

- Fixed location: programs and data can be moved between the individual sites in a distributed system. As location is no longer implicit, some explicit notion of location and change of location is required.

  fixed location → integration

- Direct binding: continuously running distributed systems must be capable of being dynamically reconfigured to accommodate changing circumstances. This requires indirect (or dynamic) binding mechanisms.

  fixed configuration → incremental change

- Homogeneous environment: there can be no guarantee that the components of a distributed system use the same implementation technology. The set of diverse technologies will change over time.

  homogeneity → heterogeneity

- Single solutions: distributed systems are created from many small, reusable distributed modules, rather than by using a monolithic approach. Typically, although not necessarily, an object-based approach is used. These modules may have very different evolutionary time scales (different life cycles); therefore the system as a whole may not exhibit the conventional life cycle phases.

  single solutions → modularisation

- Visibility: knowledge of the network and/or systems that support that service is not a prerequisite for using a supported service. Also, the location of a service may be hidden from the user.

$$\text{visibility} \rightarrow \text{transparency}$$

Following on from this, why should anyone want to create and/or use a distributed system?

Well, as already stated in the body of the text, businesses are physically distributed. Large businesses have sites spread throughout the country (or, increasingly, world). Thus, business processes are distributed; and distributed systems are the most logical way to support distributed processes.

Many of the properties that make distributed systems different from centralised systems (as listed in the last section) are also the benefits of distributed systems. Brief descriptions of the benefits follow:

- Locally encapsulated state: this means that there are fewer side-effects during processing, because access to the state is only possible via a well defined interface.

- Concurrent execution: program execution on separate processors in a distributed system takes place concurrently. This can improve the performance of the system as a whole. (The drivers for this are a desire to exploit desktop processing power and the desire to operate autonomous workgroups.)

- Fault tolerance: in the event of some part of a distributed system failing, the remainder can often continue operating without it.

- Migration: programs and data can be moved between the individual sites in a distributed system. In other words, applications and data portability.

- Incremental change: distributed systems can be dynamically reconfigured to accommodate changing circumstances.

- Heterogeneity: a distributed system can be implemented using a diverse range of technologies.

- Modularisation: distributed systems are created from many small, reusable, distributed modules, rather than by using a monolithic approach. This aids flexibility, maintainability, interoperability and reuse.

- Transparency: knowledge of the network and/or systems that support that service is not a prerequisite for using a supported service.

Distributed processing is an extension of the concept of computer networking. A simple networking system provides communications between otherwise independent autonomous computer systems to enable remote access of individual information and processing resources. In summary, distributed processing is concerned with the combination of such resources into an integrated structure to support the needs of a particular enterprise, such as the automation of information processing in an office or factory.

Networking is an important part of distributed processing, but communications alone are not sufficient to ensure the required degree of coupling between the components of a distributed system. Constraints on the structure, behaviour and interfacing of the networked components themselves are also necessary if they are to be able to support distributed processing. So much for basic principles. Now for some of the mechanisms for achieving distributed processing, along with some guidance on applications. The remainder of this appendix delves into some of the technical background to constructing distributed solutions. Although primarily focused on the system developer, much of this is vital background for purchaser and manager alike.

A2.2   THE CONCEPTS AND APPROACHES FOR MEETING THE CHALLENGE

It should be borne in mind that most of what follows is centred on software development. After all, this is the basic means of delivering processing systems. Nonetheless, the concepts and guidelines below are central to understanding distributed systems, and this goal is shared by designers, suppliers and users alike.

First we look at some of the general principles that are being adopted in developing the software for distributed computing systems. We then move on to describe a couple of the more prominent architectures into which developed software components fit. Some of the principles for developing distributed systems covered here have already been introduced in Chapter 8 in the context of network management (based, predominantly, around the concept of co-operating objects). This specific area typifies the more general approach being taken in the world of distributed processing.

The basic enablers

Flexible solutions will help enable quick and cost-effective change. Each new solution that is developed should move incrementally towards a more flexible infrastructure. The more flexible our infrastructure, the more we can exploit it.

Distributed (or, more to the point, distributable) solutions aid flexibility. These guidelines are provided to help developers to design solutions which are both flexible and distributable. The two properties are closely related.

Object orientation

The term object orientation has, over the last few years, been very broadly applied. Here we use it in its most general sense, that of abstraction, encapsulation, inheritance and polymorphism. An object is an abstract encapsulated entity which provides a well-defined service via a well-defined interface.

Object-oriented development means that the problem to be solved (business requirement) should be decomposed into component problems whose solutions can be independently useful. These component problems should then be solved, producing encapsulated solutions, taking care to abstract away from unnecessary detail. (This does not, necessarily, mean that solutions should be implemented using object-oriented programming languages; we are talking about design, not implementation.) Each individual object should be developed as a logically separate project, to ensure that they are developed to be independently useful.

The resulting solution should be comprised of many interacting 'objects', each of which is independently useful, encapsulated, well-described and has a well-defined interface. Each object can be considered to be a 'generic component'.

This form of object-oriented approach to solution development will help to enable our infrastructure to be flexible, because each object offers an independently useful service which may be utilised in a variety of situations. Also, the implementation method is more flexible because the implementation of an object may be achieved in any way and altered at any time, as long as its functionality and interface remain the same.

It is important to realise that objects need not be confined to providing purely functional services. Many requirements can, and should, be expressed as objects (e.g. business requirements, security requirements, naming requirements, management, etc.).

Reusability

Object orientation allows flexibility through reuse. Objects are inherently reusable, and so allow infrastructure changes and additions to be made quickly and cost-effectively.

Reuse can be viewed in two ways. First, static: this is the traditional view of 'reuse'. New services are provided by decomposing the problem into its component problems, satisfying as many of these as possible using generic components from a repository, then solving any remaining problems by developing new generic components, which are then added to the repository.

This is still an entirely valid approach but it must be emphasised that these generic components are not exclusively chunks of code. They are objects, components which provide a service.

Secondly, there is dynamic reuse: this is a more radical view of 'reuse'. Here components provide their services, on a dynamic basis, to other components in the infrastructure (see 'trading' and 'federation', below). A component is 'reused' every time it provides a service to another component. This approach follows the thinking that components are more useful when sitting in the infrastructure making their services readily available, rather than sitting in a repository waiting to be found. Dynamic reuse poses a considerable challenge in terms of configuration management, as the static issues encountered during integration have to be dealt with at run time.

There is a fine balance between these two forms of reuse. We must get the 'granularity' of reuse correct.

An entirely static approach to reuse would mean massive replication of components and subsequent problems with management of change and thus, limited flexibility. An entirely dynamic approach to reuse would require exceptionally reliable and incredibly high-capacity networks to cope with the volume of communication required For evolution towards a flexible infrastructure we require both forms of reuse, in the right proportions.

Remoteness

Objects are most useful, and flexible, when they are able to interwork remotely. That is to say that an object is able to interwork with another object even though their only interaction may be via a communications network. The objects need not be on the same hardware platform, nor even in the same country!

Objects which may interwork remotely aid flexibility because they may be geographically relocated without causing loss of service availability, as long as they are connected to the communications network and can be located by other objects.

It is also important to realise that a local object is simply a co-located special case of a remote object. Co-located objects should still interwork as if they were remote, to ensure that they could still interwork if one was relocated.

With the benefits of remoteness come the constraints. Remote interactions are more likely to fail than local ones, because they rely on the underlying communications. Objects must be able to tolerate failures.

Client–server

In a client–server interaction, an object provides a service to an object requiring a service. The object providing the service is known as the server and the object using the service is known as the client. This interaction need have no fixed elements. Any object may act as a client and/or a server in any number of interactions, simultaneously.

Client–server is really nothing new, it is just an explicit expression of what really happens when two objects achieve a two-way interaction. It is like having a service level agreement between the two objects. They both know their positions and what is expected of each other during their interaction. In the broadest sense, client–server can be used to describe physical arrangements (i.e. the objects may be a PC client and a mainframe server)

All objects in the infrastructure should have at least one client or server role. An object which has neither contributes to the problem of legacy. It can neither provide a useful service, nor make use of available services. It provides a one-off solution to a problem and hinders evolution towards a flexible infrastructure.

The client–server approach aids flexibility because a client can use any server offering the required service. Changes in the infrastructure are less likely to deprive objects of the services that they require because the services can be available on more than one object. Client–server also aids flexibility by facilitating interoperability, trading and federation.

To achieve maximum benefit from the client–server approach to design, care should be taken to avoid developing servers which are aimed at specific clients. Servers should be designed in the anticipation of many different clients, so that they are most flexible.

Transparency

Transparency means that, during a client–server interaction, certain properties of the interaction are hidden from the client (e.g. the location of the server providing the service, the architecture which supports the server, the route taken by the communications link). In effect, the user should see what he or she wants to use unencumbered by intervening hardware, software or network.

Transparency aids flexibility because it allows the infrastructure to be reconfigured without needing to inform objects of the resulting changes to aspects of interactions which are transparent to them.

However, certain aspects of interactions (especially those associated with quality of service) should be dealt with carefully (e.g. a client usually needs to know the cost and time required by a server to provide a particular service). These aspects of interactions should not be transparent to the client. In these cases, it must be possible for the client to declare its minimum requirements for these aspects of the interaction (see trading, below).

Interoperability

Interoperability between two objects means that they can communicate with each other. There are many levels of interoperability. I may call Japan and interoperate on a technical level (I can hear my Japanese contact) but we may not be able to interoperate on an information exchange level, because we have no common language (unless he or she speaks English).

One classification of the levels of interoperability (in ascending order of abstraction) is as follows.

Communications protocols: these simply allow objects to send and receive messages to/from each other reliably. The content of the messages is not important at this level. For example, the TCP/IP protocol.

Interaction models: these allow objects not only to pass messages to/from each other but to achieve useful interactions (useful sequences of message passing). Examples of possible interactions are a 'signal' (where a component of an object (a 'sub-object') simply sends a one-way message to another 'sub-object', within a single object), an 'interrogation' (where an object sends a request to another object, and gets a reply), an 'announcement' (where an object send a one-way message to another object) and a 'broadcast' (where an object sends a one-way message to many other objects). For example, ODP (see later) and CORBA interaction models.

Information standards: these allow the objects to understand the information being exchanged. This incorporates both the representation of the data. For example, people's names (do the initials come before or after the surname? do we use initials or full names?), and queues (FIFO or LIFO?).

Operation semantics: these allow the objects to follow the operation being performed. For example, you can use TCP/IP (communications), send an RPC (interaction) and use an agreed syntax for the information being passed (standards), but the server still needs to be able to interpret what that information means (and hence what it has to do).

Formal methods groups are working, with some success, on defining solution to these problems but at present there is no complete answer. This is a major constraint and needs further investigation. Without these mechanisms for accurately matching service requests with offers, a trader is much less effective.

Federation

Objects need to co-operate with other objects. Interoperability and trading facilitate this, technically, but there may be political barriers. It is possible that two objects will be in different 'domains of control'. The objects are controlled by different authorities.

Each authority needs to maintain control of its own local resources, and apply its own policies locally (e.g. naming, security, reliability, addressing and routing). But in order to be most effective it should make use other objects and allow itself to be used, by providing an agreed service and management interface.

Federation means that using a service controlled by another authority is not a problem. In a federated infrastructure, clients can (and should) make use of services provided by servers controlled by other authorities without needing to impose the client's policies and design constraints on the server. The two communicating objects negotiate on policies and mechanisms before they interwork. Federation is a tool for managing complexity and allowing scalability.

In effect we are recognising that objects can benefit from teamwork in the same way that people work better as a team. The ability to interlink arbitrarily allows objects to share resources and build common context in an unplanned incremental manner, much as people do.

A major constraint on federation is that same as that on trading; our current inability to unambiguously specify service. When two objects interwork, their interface agreement can be seen as service level agreement. When the two objects are in different domains of control, this interface agreement must be binding (preferably by law, or infrastructure-wide regulation). Without this type of agreement, objects cannot (and thus will not) rely on other objects to provide service. We must find a way to formally specify service level agreement between objects.

Scalability

Objects will always need to change, grow and merge. An object that provides access to the records of all of a companies sales outlets will grow (they hope!). An object that calculates tax will need to change.

One of the challenges to the designer is to separate objects of long-term applicability from those which change frequently. For example, you might separate the tax calculation engine (long term) from the objects which provide details of current tax rates, thresholds and allowances. As well as being functionally flexibly, objects must be scalable. Changes from small to large, slow to fast, specific to general-purpose, etc., should not impose fixed constraints on the extent to which objects can interwork.

Scalability of objects can be achieved by designing them using mechanisms which scale; that is, mechanisms which are efficient when solving small/slow/specific, etc. problems but are sufficiently functional to suit large/fast/general, etc. configurations.

Portability

Portability of an object refers to the ease of which it can be converted to provide its service from a different technology platform. Portability aids flexibility simply by aiding static reuse; portable objects may be easily implemented elsewhere.

In a properly object-oriented development environment, portability should come naturally. Each object should be implemented using components ('sub-objects') which each have a well-defined, useful role. A limited number of these 'sub-objects' should be responsible for interacting with the underlying technology. Thus, to port the object to a new technology platform, only a few of its components need to be changed.

This re-emphasises the need for a totally object-oriented approach to design. Anything which is independently useful should be encapsulated as an object.

Ownership

Ideally, all objects in a flexible infrastructure should be owned. Ownership provides the mechanism for attaching responsibility to objects. The owner of an object is responsible for ensuring that it can provide the services it advertises, and for dealing with change requests.

Ideally, all processes, applications ('systems') and data items should have identifiable owners. This is a valuable step forward, but it must be extended so that ownership is individually applied to all objects.

Process ownership   Processes are purely functional descriptions of how business objectives are to be achieved. A process owner is responsible for: '. . . ensuring that his or her process is operating to world-class standards and that all requirements are met both for external and internal customers . . .'.

It is already well understood and documented that the separation of requirements from solutions aids flexibility. Individual ownership of processes and the objects chosen to satisfy them allows us to separate our business requirements from their solutions.

Application ownership   For legacy systems, all components of an 'application' are owned by the unit who funded its development. However, as we evolve away from legacy towards a more flexible infrastructure, the service provided by a traditional 'application' is provided by a number of independently owned objects. Ownership of all the objects required to provide the service once provided solely by a single 'application' is unlikely. Thus the idea of 'application' ownership must evolve towards 'object' ownership.

Data ownership   A legacy system usually owns (and stores) most of the data that it processes, and the data is rarely made available externally to that system. In a sense, the data is held 'hostage' by the system. This is a problem which leads to massive data replication (and thus inconsistencies) because every legacy system needs its own copy of the data processes.

Independent ownership of data provides the mechanism to allow data to be available to any object which needs them. Data should not be implicit to the entity which processes them. Data items should be encapsulated as objects. The service provided by the object is access to the data. In this way, any object which requires the data can access them.

Development tools

There are many tools available to developers, providing support in all areas, from business analysis to detailed design. Developers should exploit tool support as fully as possible, but should take care to select only tools which support the design guidance provided in this document and those that it references. A new technology toolkit should not be chosen simply because it incorporates all the latest techniques. Fashion has no place in development!

Ideally, the chosen tools should interwork (e.g. a business analysis tool should be able to output results to a requirements specification tool). However, this is not commonly possible and there is a need to identify which tools can be brought together in this way.

It has to be said that, at present, there are few tools that assist with distributed design *per se*. There are plenty of general and support tools, but few that tackle the key issues raised by distribution.

One final piece of guidance in this section is an acknowledgement that 'green field' situations are rare if not extinct. It is almost inevitable that any system will have existing systems with which it needs to interwork. These 'legacy' systems need to be incorporated into the overall plan for any distributed system.

Usually, a legacy system is one that has been developed to fulfil a specific requirement without too much concern for how the system will evolve and need to communicate with other systems in the future.

The majority of current systems are legacy systems so we must find ways of evolving from these towards a more flexible infrastructure. This is not an issue that can be ignored, at least not safely.

There are two issues here. How do we handle today's legacy? In other words, how can we exploit the huge resource of our legacy systems while evolving towards a flexible infrastructure? In addition how can we ensure that new solutions do not become the legacy systems of tomorrow?

In a flexible infrastructure the best possible role for a legacy system is as a server for the useful services it offers. Legacy systems must be integrated so that their services can be exploited as fully and as easily as possible.

There are many possible levels of integration. At one extreme, the legacy system remains completely intact, i.e. any integration takes place within the clients that use it. At the other extreme, the legacy system can be completely re-engineered to conform to the flexibility guidelines. There are many points on the spectrum between these extremes; the following are just a few examples.

- Client makes right: there is no integration of legacy system, you just have to arrange any clients to work with your legacy systems. This is a restrictive (but clean) solution.

- Proxy interfaces: the legacy system front end is placed in the server. The client end is simplified.

- Structured proxy access: the legacy system is re-engineered to improve proxy access. This relies on getting information from the legacy systems as the interface to the legacy. No direct interaction takes place.

- Shrink-wrapped legacies: this is full integration of legacy systems via well-defined interfaces using a trader to control client–server sessions.

The extent to which integration is pursued depends upon factors such as the cost of re-engineering, the risk of re-engineering, the level of access to systems (some may be third party systems) and the importance of the service. In practice, legacy systems usually pose a headache. The guidelines given above are intended to avoid (or at least minimise) the problem of interworking with tomorrow's legacy systems.

A2.3   THE GUIDING ARCHITECTURES

The main reason for dwelling on the general concepts (that seem to abound) in distributed processing has been to put a complex and diverse subject into some context. The guiding frameworks that describe distributed systems are similarly complex, a full-time occupation for many technocrats. Having said this, there have been a number of initiatives aimed at making the idea of distributed systems into a reality. The routes to date have been based on 'open systems', specifying sufficient criteria for systems from different suppliers to interwork (at least) and to be interchangeable (ideally).

Some initiatives such as ODP have been started by standards bodies, whereas others, such as the Open group, are vendor driven.[1] The overall aim is common: to define the standards and frameworks for interoperability of computer systems. This, in turn, fulfils the aim of allowing systems to be assembled with interchangeable hardware and software components options to best suit the end user's particular needs. The quest for viable open systems has gained momentum since these initiatives were launched, and there are now usable open system components that conform to well-publicised standards.

To complete this appendix, we explain two of the main sources of distributed processing standards: Open distributed processing (ODP) and distributed computing environment (DCE). The former is standards-driven, the latter an industry initiative. These are by no means the only important players involved in open systems. Others (the object management group (OMG), for instance, which is driving CORBA) are providing useful parts of the overall jigsaw. Some idea of the range of activities and proponents can be derived from Figure A2.4, which shows the likely evolution path of some of the key elements of distributed computing.

Some, but not all, of the above elements will be explained below. The main point of the figure is to highlight the fact that distributed computing is far from settled. Following the above, the remainder of this section will clearly not define how a distributed system should be built. The specifics of user access, application processing and data management will vary according to circumstance. ODP and DCE do, however, provide a framework and components, respectively.

Open distributed processing (ODP)

In recent years, the ISO has been working on a framework for open distributed systems, called the reference model of open distributed processing

---

[1] This is an industry standard consortium that seeks to publish detailed system specifications for open standards. By way of illustration, they 'own' the Unix trademark, thereby bringing focus to its various flavours (Ultrix, AIX, Solaris, etc.).

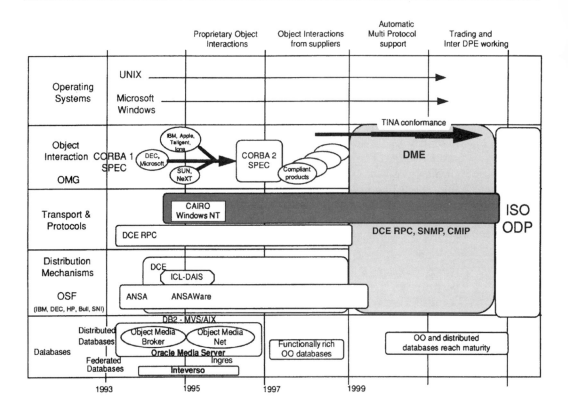

Figure A2.4   Evolution in distributed computing

(RM-ODP). The results are to be published jointly with CCITT. This section describes first the main structuring principles described in the RM-ODP (often referred to as simply ODP), then the subject matter of the five parts of the RM-ODP standard.

In the late 1980s it became obvious that there were two parallel groups developing open systems standards: in the ISO open systems interconnection (OSI) communications standards were well advanced; and in the IEEE the POSIX system services interface standards were being defined. (POSIX is the portable operating systems interfacing for Unix. It is part of the IEEE standards and is concerned with standardisation of the application programming interfaces (APIs) that allow applications that can be ported from one machine to another to be written.) These reflected two communities with two preoccupations: communications experts wanted to allow systems to interconnect by defining open communications standards, and systems software experts wanted to make software more portable by defining open operating system standards. The ISO responded by establishing work on the RM-ODP, whose goal is to provide the basis for open distributed systems.

The RM-ODP sets the context for a new family of open systems standards whose objectives include system interconnection, software portability and applications interoperability.

The ODP framework

A key success of early work on the RM-ODP was the separation of the concepts and language used to describe distributed systems, from the prescriptions to be applied to ODP-conformant systems. The result was the development of the RM-ODP descriptive model, a general framework within which to describe any complex system: a new design, an established system, an open system, a proprietary one, a fully-distributed system, a centralised one, and so on.

The descriptive model is of immediate benefit: one barrier to the integration of existing information systems within large organisations is that their different designs, languages, assumptions, requirement, and purposes are often far from clearly described. Such descriptions as do exist all start from different baselines. Relating all such descriptions to a common framework helps one to understand the possibilities and difficulties of interpretation.

The descriptive model comprises two key elements: a framework for system description (the viewpoints) and a system structuring mechanism (the object model).

The ODP viewpoints

To deal with the full complexity of a distributed system, the system can be considered from different viewpoints. Each viewpoint represents a different abstraction of the original distributed system. By partitioning the concerns to be addressed when describing all facets of a system, the task is made simpler.

Informally, a viewpoint leads to a representation of the system with emphasis on a specific concern. More formally, the resulting representation is an abstraction of the system; that is, a description which recognises some distinctions (those relevant to the concern) and ignores others (those not relevant to the concern). There are in fact a number of possible abstractions. Of those thought to be more useful for the description, analysis and synthesis of complex systems, some are concerned with objectives, others with realisation, and still others with behavioural attributes. The viewpoints each address the whole of a given distributed system, and are as follows.

Enterprise viewpoint The enterprise viewpoint provides members of an enterprise, in which the information processing system is to operate, with a description showing how and where the system is placed within the enterprise. The model from this viewpoint captures business requirements and objectives that justify and orientate the design of the system, and can be expressed in terms of objects representing user roles, business and management policies, the system and its environment.

Information viewpoint Information managers, information engineers and business analysts look at systems from the information viewpoint. From this viewpoint, parts of the system that are to be automated are not differentiated from those performed manually. Information structures and flow are modelled, and rules and constraints that govern the manipulation of information are identified.

Computational viewpoint Application designers and programmers think about systems from the computational viewpoint. From this viewpoint, programming functions and data types are visible. The structuring of applications in this viewpoint is independent of the computer systems and networks on which they run; requirements for distribution transparency are identified here.

Engineering viewpoint Operating systems, communications and systems software experts look at systems from the engineering viewpoint. The objects visible from the engineering viewpoint are control and transparency mechanisms, processors, memory and communications networks that together enable the distribution of programs and data. It is concerned with the provision and assurance of desired characteristics such as performance, dependability and distribution transparency.

Technology viewpoint Those responsible for the configuration, installation and maintenance of the hardware and software of a distributed system are the primary users of this viewpoint. Concerns include the hardware and software that comprise the local operating systems, the input–output devices, storage, points of access to communications, etc.

A reference model is not simply a descriptive tool. It establishes a style and basis for system design and implementation, allowing the use of common components and common design methods and representations; in short, an architecture. However, the ODP architecture deliberately focuses only on those reference points where conformance will be needed, and minimises unnecessary constraints.

DCE

Distributed computing environment (DCE) is an industry-led initiative. It comprises a set of services and tools that have been defined to support the creation, use and maintenance of distributed applications in a heterogeneous computing environment. DCE is produced as a technology by the Open Software Foundation (OSF), but will be supplied in product form by value-added retailers, typically the suppliers of computing platforms. DCE is now becoming commercially available, and support for it is growing. (For instance, IBM are supplying the DCE product on their RS6000 machines, and DCE for Sun SPARC machines is supplied by Transarc Corp. Software to allow PCs running Microsoft Windows to be DCE Client machines will be supplied by Gradient Technologies.

To recap on earlier points, distributed systems bring a number of benefits over centralised systems.

- Functional: special-purpose hardware or software does not need to be duplicated on every node that needs access to it.

- Economic: it may be more cost-effective to have many small computers working together than one large one (realising the benefits of right-sizing). This is also a more flexible configuration.

- Reliability: reliability and availability are enhanced as a result of replication of data and functionality.

The overriding point, though, is that distributed systems are a natural match to the way that more and more people are working. This is becoming ever more the case. To quote H. A. Simon, 'The question is not whether we decentralise but how far we decentralise', and distributed systems provide the flexibility to support this business imperative.

There are some drawbacks, however: there is the management overhead of large distributed systems, covering network, service and information management; and currently there is little practical experience of developing and maintaining distributed applications.

DCE's set of services is integrated and comprehensive. They use one another's services whenever possible, since many of the DCE components are themselves distributed services. In addition to supporting the development of distributed applications, DCE includes services that address some of the new problems inherent in the distributed system itself, such as data consistency and clock synchronisation. DCE provides a basic set of tools (control programs) to configure and maintain the main DCE Servers. They are not aimed at system management of a large distributed environment; the distributed management environment will eventually provide this (albeit for enough into the future to be something to rely on). DCE provides interoperability and portability across hetero-geneous platforms. The DCE architecture operates over different operating systems, network protocols and hardware platforms. Using DCE, a process on one computer can interoperate with a process on a second computer, even when the two computers have different hardware and operating systems.

Why use DCE?

Currently, two important uses have been identified for DCE. First, the need to make more effective use of resources distributed across a network. In doing

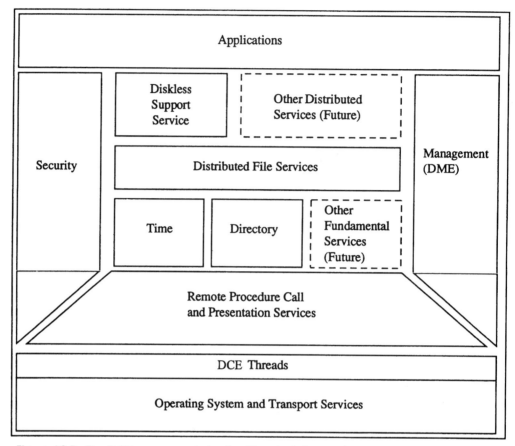

Figure A2.5   The DCE components (or the 'Theatre' diagram)

so, it addresses the minimum requirement of a distributed system which is to make the resources look, and behave as though they were one system rather than a collection of systems. Secondly, it can be used as a platform across which to integrate operational environments (for example, PC, UNIX and mainframe). This obviously ties in with the first point and addresses issues on interoperability, portability and reuse.

DCE is an architecture that will support fundamental services, a solid base for commercial enhancements. For example, Encina is a transaction processing (TP) monitor based on DCE components In turn, the IBM CICS/6000 product is based on Encina.

DCE architecture

DCE provides an integrated and comprehensive set of tools and services to support distributed applications. It is independent of operating system and network. Figure A2.5 shows the components of DCE.

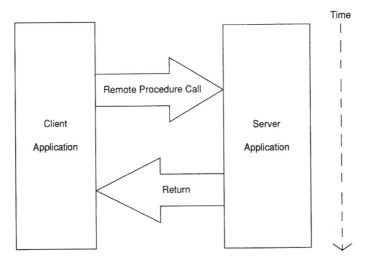

Figure A2.6 RPC operation

Each of the components shown in the diagram has a specific function. We now briefly explain the main ones.

DCE remote procedure call

The model for the remote procedure call (RPC) is similar to a local procedure, or function call, in a high-level programming language. The calling program (client) passes several parameters to the called function (server). Results are either returned as output parameters, a return value or both.

This is a synchronous operation (the client is blocked until the result is returned) and is illustrated in Figure A2.6.

For a client to locate the server, the server must register its interface with a directory service. The registration operation that the server invokes is known as exporting the interface. The locating operation that the client does to check the addressing information is known as importing the interface.

Once the location is known, it must still bind to the server before any remote procedure can be called. Different binding techniques can be used, depending on the needs of the application. It might

- bind before the first call

- bind at the time of the first call

- rebind for each call

- rebind automatically to another server in the event of a server crash.

DCE RPC consists of an interface definition language (IDL) and a runtime service (RTS). The IDL has a compiler that automatically generates code to

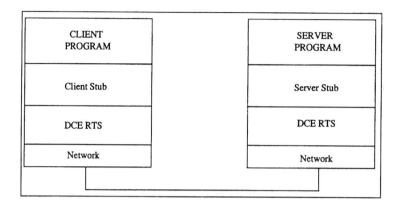

Figure A2.7   RPC facilities

transform procedure calls into network messages. The RTS implements the network protocols by which the client and server sides of an application communicate.

IDL is a tool that hides the complexities of the multiprotocol support of DCE RPCs for application programmers. It is used to specify the interfaces of the procedures that are to run remotely. The IDL compiler uses the IDL code to produce a header file, as well as stubs for both the client and server programs. The stubs invoke the RTS call necessary for transferring and converting data to the desired representation. This is illustrated in Figure A2.7.

The RTS has three major components

- communication services: these services access the network and allow simultaneous execution of different transport and RPC protocols

- naming services: these access the distributed system's directory service to locate servers by specifying a given interface or an object.

- management services: these manage RPC services locally or remotely.

DCE threads

Threads is a mechanism that allows several flows of execution to take place concurrently through a program. With DCE, it is implemented as a set of library routines that enable the development of concurrent applications. These are known as application level threads, which can be mapped onto lower-level kernel threads if provided by the operating system. DCE threads are based on the POSIX 1003.4a thread standard. Threads bring potential performance advantages through concurrent execution of code, especially on multi-processor machines. In the client/server model, concurrency helps to improve the performance of the server. Typically, one server thread will listen to incoming requests and then despatch them to other threads to handle (serve).

DCE security service

The DCE security service provides controlled access to resources in the distributed system. There are three aspects to DCE security: authentication, secure communication and authorisation. The identity of a DCE user or service is authenticated by the authentication service. Communication is protected by the integration of DCE RPC with the security service. Finally, access to resources is controlled by comparing the credentials conferred to a user by the privilege service with the rights to the resource which are specified in the resources access control list. The login facility initialises a user's security environment, and the registry service manages the information (such as user accounts) in the DCE security database.

DCE directory service

The directory service provides a way to name and locate objects within the distributed system. It is made up of two main parts: the cell directory service (CDS) manages information about resources in a group of machines called a cell (for example, the machines on an LAN); the global directory service (GDS) implements an international standard directory service based on the X.500 standard (CCITT standard, issued by the Internation Telecommunications Union (Geneva). It is used by applications to locate objects outside the local cell. This can mean anywhere in a global network.

Both CDS and GDS can be accessed using a single API, the X/Open Directory Service (XDS) API. Using this API, applications can be built that are independent of the underlying directory service.

DCE distributed time service

This service is used to provide a consistent time frame across a distributed environment, which could potentially be based in a global network spanning several time zones. If local system times are used, this will cause inconsistent time stamps, which will affect distributed applications.

This last point is one that affects both machines and people. Co-operation across time zones can accentuate the problems of working in a distributed environment. The problem has a human as well as a computing side. To complete the picture of distributed computing, we dwell briefly on the people side: computer-supported co-operative working (CSCW), a subject that has come to be studied under the popular banner of 'groupware'.

A2.4   GROUPWARE—THE PEOPLE SIDE OF DISTRIBUTED PROCESSING

Groupware is defined to consist of computer-based systems that support groups of people engaged in a common task (or goal) and that provide an interface to a shared environment. The notions of a common task and a shared environment are crucial to this definition. This excludes multiuser systems, such as time-sharing systems, whose users may not share a common task. The definition does not specify that the users be active simultaneously either, so 'follow the sun' applications (as introduced in Chapter 2) can be considered as groupware examples.

There is no rigid dividing line between systems that are considered groupware and those that are not. Since systems support common tasks and shared environments to varying degrees, it is appropriate to think of a groupware spectrum with different systems at different points on the spectrum.

A conventional timesharing system supports many users concurrently performing their separate and independent tasks. Since they are not working in a tightly coupled mode on a common task, this system is usually low on the groupware spectrum.

In contrast, consider a software review system that electronically allows a group of designers to evaluate a software module during a real-time interaction. This system assists people who are focusing on the same specific task at the same time, and who are closely interacting. It is high on the groupware spectrum.

There are two taxonomies useful to view the variety of groupware. These are as follows.

Time–space taxonomy: groupware can be conceived to help a face-to-face group, or a group that is distributed over many locations. Furthermore a groupware system can be conceived to enhance communication and collaboration within a real-time interaction, or an asynchronous non-real-time interaction. These time and space considerations suggest the four categories of groupware that can represented by the 2×2 matrix shown in Figure A2.8.

Meeting room technology would be within the upper left cell; a real-time document editor within the lower left cell; a physical bulletin board within the upper right cell; and an electronic mail system within the lower right cell. A comprehensive groupware system would serve the needs of all of the quadrants.

Application-level taxonomy: This taxonomy is not comprehensive, and many of the listed categories overlap. It is intended primarily to give a general idea of the breadth of technology applications that apply to the groupware domain:

- message systems

- multiuser editors

- group decision support systems and electronic meeting rooms

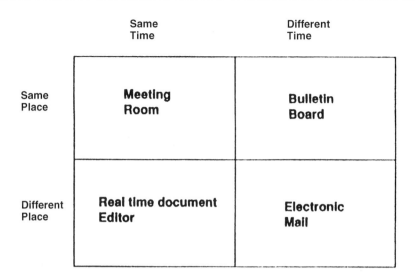

Figure A2.8   Categories of groupware

- computer conferencing
- tele- and video-conferencing
- intelligent agents
- co-ordination systems

To give some feel for the sort of work that has gone on in this area, we now give some examples that highlight the range of technology and related applications.

GROVE (group authoring software)

The GROVE system was developed at MCC, Texas, in 1988. The need was for several researchers to write a joint paper under considerable time pressures. This came together with an objective to explore 'distributed meetings', where the participants are not in the same place at the same time.

Calendar management

In a study by Grudin the track record of electronic calendars was analysed to show how the track record of many of the less successful distributed working applications can be traced to five common factors

- if the people who have to do additional work are not the people who perceive direct benefit from use of an application
- if the technology threatens existing political structures or certain key individuals

- if it does not allow for the wide range of exception handling and improvisation that characterises much group activity
- the complexity of the applications makes it very difficult to learn and generalise from our past experiences
- our intuitions are especially poor for multiuser applications.

These key indicators provide a useful checklist against which the applications of the Information Age can be measured. Several subsequent studies have looked at the effectiveness of tools such as Lotus Notes, PC Anywhere and Windows for Workgroups, all of which serve some useful purpose but do not supplant the need for a disciplined approach to group working.

The CO-ORDINATOR

This is probably the best known CSCW E-Mail application. Its function is 'conversation management'. The CO-ORDINATOR was designed as a computer-based electronic mail system to facilitate exchange, clarification, and negotiation of commitment in organisations. This demonstrated the importance of having informal channels that the distributed team can use to exchange background, non-urgent and supplementary information.

Work team support—the Florence project

Four researchers from the University of Oslo set out to '. . . explore whether—and how—computer systems can be used in nurses' daily work rather than focusing on more administrative work tasks, carried out by head nurses and the like'. The nurses' information function was complex. It involved at least passing all relevant information to the next shift, co-ordinating information with other nurses to have an overview of the state of the ward, arranging for different medical doctors to meet (or conveying information to them individually), and relaying the doctors' daily programmes for each patient back to the ward.

All these meetings cut into the time available for patient care. The CSCW-supported competence would be a way of 'packaging variety' so that everyone got the necessary information without all the meetings.

Computer-supported spontaneous interaction: CRUISER

Social interactions and resulting personal relationships are critical to the success and effectiveness of the organisation: they foster the development of mutual understanding, define the channels through which information flows into and throughout the organisation, and increase the effectiveness of communications among technical employees. Workplace social relationships are also known to be important elements of job satisfaction and sources of social support. This raises the question of how to most effectively support work between colleagues and collaborators when they are necessarily separated by large offices or different workplaces. CRUISER attempts to tackle this problem. It takes its name from 'cruising', the teenage practice of the 1950s of piling into someone's car to visit coffee bars in search of social encounters!

Technically, the design assumes the availability of desktop full-motion video communications, high-quality full duplex audio, a switched multimedia network under the control of a local computer, and integration of video images and computer-generated graphics: the very stuff covered in the main text.

To complete this appendix, we give a set of guidelines for organising a distributed team. Each of the following points is no more than commonsense, but they do represent a distillation of the key results of research and practical experience. Although simple, the following are fundamental to effective working in the Information Age—some of the new rules of the game.

High-quality planning

This is a crucial part of successful distributed working. It is essential to establish at the outset who is responsible for what and how they communicate. A level of formality needs to be put in place as the visibility and immediacy of working side by side is removed. Issues such as problem resolution should be given high priority.

Fallback options

Network reliability is of paramount importance as it is the basic working medium. But even the best networks fail sometimes, so a readily invoked fallback mechanism must be in place (e.g. a low speed link over telephone lines instead of a high-speed data link). The fallback arrangements should allow operations to carry on (albeit non-optimally), especially when working across time zones, as resynchronisation can cost dear

Monitored file transfer

It should never be assumed that items arrive as sent (indeed, at all). Content and arrival checks should be part of the operating procedures.

Review of documentation

The sensitivity of remote readers to minor mistakes, ambiguities and inconsistencies in received information can cause considerable delay. Important information passed from one place to another should be peer-reviewed prior to sending.

Allocate multiple tasks

The inevitable delays in resynchronising operations across time zones and the certainty that no operation goes without a hitch mean that local fallback tasks should be reserved. Work can then continue while the mainstream blockage is resolved.

Prioritise communications

Messages between teams at different locations should be prioritised so that the important and urgent issues can be partitioned from the interesting and background.

Be prepared for cultural differences

There are considerable differences in approach from place to place (e.g. command of chosen language, willingness to work unsociable hours, level of expected formality in work practices, basic skill levels, etc). These have to be accounted for in plans.

Ensure the human touch

Encourage team members to exchange personal information: user groups for casual messages, a picture board, etc. This should be one of the overheads planned in at the start of the project.

Beware time and date stamps

When time differences exist between teams, date and time stamps on files and documents cannot be trusted. A wise precaution is to set all machines onto a common clock; this can be usefully in avoiding the situation where work is rejected, as it was completed before the current local time!

Allow redundancy

Teams in Sydney and London will not work at twice the rate of one team, even though they can put in twice as many hours in a day as a single team. Some operational overheads must be allowed for extra formality and for recovery from problems. An elapsed time gain of 80% rather than 100% would be reasonable here.

Much of the above is fairly straightforward and can readily be built into the distributed applications and systems that are deployed. The important point here is that the Information Age will not become a reality simply through technology. There will have to be an understanding of how people work as virtual teams and processes to support this will have to be put into place. Commercial success will depend on this as much as it does on the technology

REFERENCES

The following set of references is general, rather than specific. The references are grouped into categories of distributed teams and distributed computing, but are not placed specifically within the text of the appendix.

Distributed teams

Abel, M. (1990) *Experiences in an Exploratory Distributed Organization.* Prentice-Hall.

Checkland, P. (1981) *Systems Thinking, Systems Practice.* John Wiley & Sons.

Grudin (1990) *Groupware and Cooperative Work: Problems and Prospects.* McGraw-Hill.

Johansen (1988) *Groupware—Computer Support for Business Teams.* Free Press.

Posner and Baecker (1992) *How People Write Together.* Prentice-Hall.

Distributed computing

ANSA. *The Advanced Network Systems Architecture Manual.* See Glossary entry for ANSA.

Birrell, A. and Nelson, B. (1984) Implementing remote procedure calls. *ACM Transactions on Computer Systems*, **2**, 39–59.

Birrell, A., Lampson, B., Needham, R. and Schroeder, M. (1986) A global authentication service without global trust. *Proceedings of the IEEE Security and Privacy Conference*, Oakland, CA, pp. 223–233.

Black, A. (1985) Supporting distributed applications. *ACM Operating Systems Review*, **19**, 181–193.

Lampson, B. (1983) Hints for computer design system. *ACM Operating Systems Review*, **17**, 33–48.

Norris, M. and Winton, N. (1996) *Energize the Network: Distributed Computing Explained.* Addison Wesley Longman.

Saltzer, J., Reed, D. and Clark, D. (1984) End-to-end arguments in system design. *ACM Transactions on Computer Systems*, **2**, 277–288.

Appendix 3

The Future of Network Supply and Operation

Any sufficiently advanced technology is indistinguishable from magic

Arthur C. Clarke

There are a number of important players who will determine the speed and penetration of total area networking. Most attention so far has been focused on the end user: the individuals and organisations that will benefit from a raft of new technology. The end user will be a major influence on the future of networking in that he or she will seek to take advantage, but there are other key players.

One that cannot be overlooked is the public network operator. Its very survival depends on providing services, capacity and management to meet user demand. Furthermore, the scale and complexity of their offerings obliges them to plan well in advance of actual demand. In many ways, it is assumed that the network operators will provide connectivity, as required. Looking back to the preceding appendix, it is taken for granted that distributed networks will be adequately connected. In practical networks, such as the Internet, there are many options for connecting to the network. These vary from the simple modem link through to a high-speed digital connection. Either way, the facility is assumed to be there and the end user is usually looking to make the most of what is on offer.

This appendix is all about the evolving shape of the network from the operator's perspective. Most of what follows is concerned with technical developments, and much of it has already been touched upon in the body of this text. Even so, the viewpoint here is that of a network operator planning to meet predicted needs, rather than that of a user planning to capitalise on new capabilities.

This foresight of the evolving infrastructure of the network is valuable in that it reinforces notions of what is possible by providing greater understanding of the advances in network technology. Before going into this, though, we dwell for a while on the equally important changes that are likely in the provision and ownership of networks.

The key point to bring out before going into any detail is that the traditional boundaries between different types of network and services will become blurred. In some cases, they may disappear altogether. For instance, not very long ago there was voice communications (telephony) and data communications. The two tended to be treated as separate entities with separate management and provisioning arrangements, even when carried on a common link.

Over the last few years the distinction between the two has faded with the introduction of terminal equipment that brings together voice, data and video services. Indeed the watchword for the future is multimedia. A time-honoured distinction is being eroded. If present trends continue, it will disappear altogether and there will be one multiservice network rather than separate voice and data facilities.

Just as the services that are carried are converging, so are the networks that carry them. The differentiation between private and public networks is no longer a simple matter of the user owning and operating the former or buying managed capacity on the latter. There is now a whole range of options to move between traditional private networks, managed network services and a private network with outsourced management. There is a whole spectrum of solutions for solving an organisation's network needs. For example, a private network may be appropriate for operations in one country, with managed network services providing the majority of the organisation's needs in another.

As far as the user is concerned, the whole network should look like one resource; the managed service that has been outsourced provides an extension of the private network. The term virtual private network is now commonly used to reflect the fact that issues of ownership and usage (as shown in Figure A3.1) need to be considered in meeting communication needs. It is likely that the increasing complexity of network and services will make virtual networks the rule rather than the exception. As technology becomes indistinguishable from magic, so the ability (and, indeed, desire) of any one organisation to own and manage all of its communication infrastructure will diminish.

The major reason behind the shifts that are shown in the figure is the advance in technology. This advance motivates (and will continue to motivate) a re-evaluation of the way in which networks are handled. For instance, it is the increasing costs and rate of change in network service provision that are likely to fuel a swing towards specialist network service providers. Furthermore, it is not just the networks themselves that reflect new technology. The structure and organisation of its users are also affected, with information rather than manufactured goods becoming the currency of growth industries. The consequences of this are to be seen in the push for

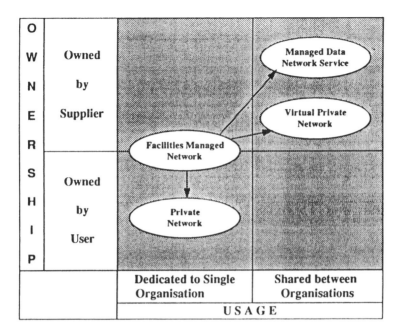

Figure A3.1   Ownership trends

globalisation, teleworking and virtual teams in many high-tech companies.

The next section looks at some of the pressures and drivers that are influencing the way in which communications are provided and managed. The remainder of the chapter considers the way in which evolving technology is likely to move us from the here and now towards the network of the future. History points the way to an important lesson in all this; that an awareness of where technology is leading is a key part of preparing for, and capitalising on, the future.

A3.1   THE FORCES THAT WILL SHAPE THE FUTURE

Over the next decade the way that communication networks are designed, provided, operated and managed will change quite dramatically. These changes will create both opportunities and threats for users, network operators and equipment/service providers alike. This section looks at the factors that will most influence these changes, with particular focus on those factors that will drive the infrastructure that the public operators will provide.

The existing network designs, technologies, operation and management procedures are under pressure from three main areas: customer expectations and requirements, increasing competition and globalisation, and the falling cost and proliferation of network implementation and service delivery technologies. We now look at each one in turn.

Customer pull

The potential market for communications bandwidth is increasing at an astronomical rate, with the increasing penetration of computer and information technology into our everyday lives. Not only is the demand from current users increasing rapidly, but there is an even more rapidly growing latent demand, which appears to be capable of absorbing any conceivable increases in bandwidth which could be made available, provided that the service can be delivered at an acceptable price. Organisations want to extend the facilities they have locally (LAN facilities) to their entire operation. Individuals want to have services (voice, data and video) delivered, on demand, to their houses.

As the network technology and the services it provides become ever more complex and difficult to manage, so the enticed user's expectations of service quality rise daily. This is not too surprising since his or her perception of the complexity of their service is confined to what can be seen on the desk. With the advent of such hi-tech yet reliable equipment as the PC on every desk, the non-availability of any service becomes unacceptable. Compounding this factor is the ever-increasing business value and perishability of information. On the international money market, loss of communication for a few seconds at the wrong time can cost literally millions of dollars.

It is worth noting that the global telecommunications network is by a large margin the most complex single entity ever assembled by man. In spite of this, it is expected to (and usually does) perform almost flawlessly.

This perishability of information, plus increasing personal mobility and demand for convenience, has also fuelled an explosive demand for mobile communications, and this will surely continue.

Finally, the globalisation of business operations means that business customers are increasingly viewing communications networks as a global resource, needing to be managed and operated on a global basis.

The key point that brings together all of the above is that an increasing number of employees are becoming 'information workers' and a growing proportion of companies rely on them as their main revenue generators. The needs of the information worker (fast, easy access to data that are distributed all over the globe) are paramount in what the future network will have to provide.

Competition and regulation

The increasing deregulation of the global telecommunications market is giving customers more freedom to choose between alternative service providers. Responsiveness to changes in customer demand for bandwidth, or new services will become a key factor in supplier differentiation. The same will be true in respect of ability to respond to demands for higher level services and flexible dynamic restructuring of a national or global customers communications requirements.

The globalisation of business operations mentioned above is also leading to the globalisation of network and communications service providers. This has (at least in most open markets) reached the point where no regional carrier or network operator can be considered immune from competition, even in its own home territory. In practical terms, all serious network suppliers, operators and providers are now obliged to operate on a global basis.

No longer will network provision be the reserve of the indigenous carrier. Suppliers with the capability of combining bandwidth plus services plus management will be able to transcend national barriers to provide virtual networks that are customised to a user's needs, irrespective of the source of the basic network components.

Technology push

The proliferation of possible delivery technologies, with the availability of optical fibre, cable television, satellite, and lower cost radio technologies has dramatically reduced the capital cost of entry to the business of delivering communications service to the end users. Combined with the globalisation of the communications network and service providers and world wide trend towards deregulation, this has meant that there is a dramatic increase in the number of potential suppliers. Differentiation will be through flexibility and quality of service.

The cost of high-volume transmission bandwidth has fallen dramatically with the advent of optical fibre technology. The historical trend of switching costs falling faster than transmission costs has rapidly reversed over the past decade. This change, along with the dramatic increase in latent demand for bandwidth arising from a heavily information-based society will dramatically reshape telecommunication networks for the 21st century.

The real costs of radio transmission have also fallen substantially for lower bandwidths, and advances in compression and processing technology have also brought about significant improvements in spectrum usage efficiency. New spectrum bands are also being opened up by the falling costs of the various available technologies, which will push the upper limit for cost-effective radio transmission close to 100 GHz in the not too distant future.

Other free-space communication techniques, such as infrared, will also make some contribution and provide a similar function to that of current radio systems, particularly for point-to-point operation.

A3.2   THE HERE AND NOW

All the above drivers are operating on an already diverse, complex and distributed base. The network of today has to evolve at every level, from its local distribution provision to its central control systems. Nonetheless, a lot of the basic elements are there to be built on.

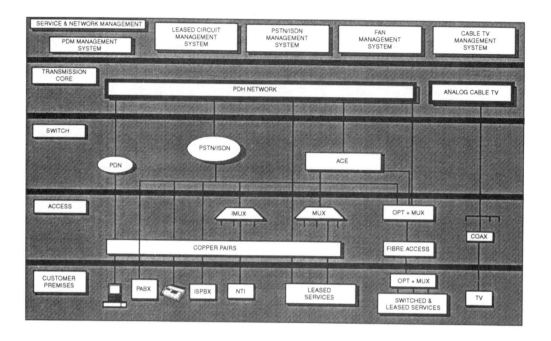

Figure A3.2   Network evolution—starting position

Figure A3.2 gives a simplified diagram of a typical high-level network architecture for a public network operator. For clarity, mobile and satellite services have been excluded from the diagram, as has the use of radio systems in the access layer.

Figure A3.2 provides the basic reference for explaining future network evolution. But first, let us define the structure of our network.

The access layer is used to deliver the service to the end customer, and is currently primarily provided by twisted pair copper. There is some use of optical fibre for higher volume business customers, and coaxial cable is used for TV distribution. There is some use of multiplexing in the access layer, but the primary interface to the switching and core transmission layers is through distribution frames where physical cross-connections are made between copper pairs.

The core transmission of the network is provided primarily by digital transmission systems, which are based on standards developed initially for the carriage of digital voice. This has an inherently inflexible multiplexing system, and is not optimised for flexible partitioning of bandwidth between the various classes of traffic that will be generated by the emerging applications which will dominate bandwidth utilisation in the coming decades.

The switch layer is used to provide the various switched services available, which are currently limited to a maximum bit rate of 64 kbit/s. There

are three separate major switching services, circuit-switched PSTN and
ISDN, which are partially integrated, and the packet-switched PDN. The
switch layer is also used to provide cross-connect functions for some leased
services.

Controlling all this is a plethora of various network management systems,
usually one for each of the different types of equipment in the network. In
addition, various service management systems are in place to control the
different services provided. Service provisioning process is still dominated
by the need to make physical interconnections at various points in the
network, which makes an automated end-to-end process for service provision
and rearrangement slow to effect and difficult to achieve with any measure of
consistency.

The basic drive is to provide what the user wants, faster, cheaper and with
greater flexibility. The drivers detailed earlier will push suppliers into
applying and controlling new technology to satisfy these needs and the
'information worker' will be one of the main judges of which operators
survive. Again, the trend towards increased competition is likely to push
virtual networks as the most practical option for capitalising on future
offerings. Already, many large companies have decided that networks are not
their core business. They are using specialists to provide and manage their
communications infrastructure.

A3.3   THE NETWORK OF THE FUTURE

The complexity and variety of telecommunication networks and services has
now grown to the point where they are becoming very difficult to plan,
control and manage using the traditional piecemeal approach. The legacy of
separate networks (for instance, for voice and data) has been mentioned
already. The piecemeal aproach has naturally evolved in response to the
various different types of service that have grown up over the last few
decades. It is now beginning to creak and will, with increasing pressure from
the forces for change, cease to be viable for all but the most straightforward of
networks or services.

Although it is relatively straightforward to manage a switched telephony
network, with only one type of service and a relatively undemanding stable
customer base, managing a global service to a major business customer with
hundreds of sites using services ranging from simple telephones through
ISDN, X.25, Frame Relay IP applications and videoconferencing, operating
through various types and vintages of equipment is quite another matter.

To meet this challenge, the telecommunications network of the future will
need to be dramatically reshaped. What is more, this future is not very far
away. To deliver an appropriate infrastructure, the picture of the public
network painted here for the year 2000 should be being built now. With the
current telephone network the largest ever man-made artefact, a comparable
data network will not be built in a day. Many operators are already effecting

evolution with overlay networks and global network and service management offerings. The end of the 20th century is likely see this progress down this path gather pace and accelerate with user demand for total area networks.

The only way to respond rapidly to shifts in the location of bandwidth demand, requirements for provision of new services, restoration of failures and changes in the characteristics of user traffic will be to establish a universal high-capacity broadband network. There will be a number of key characteristics of this network

Flexible access networks will be implemented to allow the delivery of a wide range of high-bandwidth services to a customer over a single managed access link. This will allow any services a customer needs to be established quickly. The dimensioning of the access links will need to allow the connection of a new service, on demand.

The mobile network will be significantly expanded in capacity to cater for the increasing expectation that it will be possible to communicate critical information to a particular individual at any time irrespective of their location.

Greater intelligence will need to be deployed in the network to allow the provision of new services, such as flexible redirection of calls according to staff locations and the customer calling patterns of each business.

Improvements in network management systems will be needed to simplify the task of managing the diversity that will be an inevitable aspect of global networks, and to speed the identification, by-passing and correction of faults in the network.

To meet customer demand for more responsive service provision and reconfiguration, all of this will need to be overlaid with more sophisticated automated service management systems that will allow the connection of a global service by means of a single transaction at any point in the network

Much of the technology and knowhow to bring this about already exists. There are already some instances of networks that meet some of these features; the future is already here for a selected few. Inertia will slow general availability, though. The sheer size of currently installed networks means that legacy systems will, to a large extent, dictate the pace of progress. Nonetheless, the way forward is set, some pioneers have braved the new frontier and the move to higher ground will provide a competitive edge for those who migrate their operations there.

A3.4 ACCESS—THE FLEXIBLE DELIVERY

The primary focus of access network evolution over the next decade will be on increasing the efficiency with which the existing copper pair distribution system can be used, while expanding capacity and flexibility through the introduction of optical fibre technology.

There will also be increased use of multiple technologies for service delivery as the viable technology options multiply and competitive pressures demand ever more cost-effective delivery. However, copper seems likely to remain the dominant form of delivery to the wall socket in domestic premises at least well into the first decade of the next century. The early introduction of ADSL and HDSL technology will allow the delivery of wider bandwidth services over the existing copper pair infrastructure. This will be supplemented by increasing use of point to point radio links for delivery of 2 and 34 Mbit/s services. The extent to which existing assets (i.e. copper pairs) can be made to 'sweat' will determine the rate at which fibre penetrates this market sector.

There is already significant use of fibre to service large business customers. This will increasingly penetrate to small and medium-sized business customers with the deployment of passive optical technology. This allows economical distribution through passive multipoint optical fibre networks with optical splitters and time division multiplexing, allowing flexible allocation of bandwidth to a number of customers on a single multipoint network.

The extension of fibre technology into the access network will allow wider bandwidth services to be delivered to the home. This will radically alter the economics of switching network design, encouraging the centralisation of major local exchange equipment and the deployment of small switching concentrators or intelligent service distribution points within the access network.

This will lead to the possibility of integrated video, data and telephony distribution through the same network, as well as multipoint radio distribution from concentrators to provide either mobile or cordless service. Radio will also find some application in providing a service to rural customers, and in providing an ability to react rapidly to changing service demand in particular areas.

By the end of the century, wavelength division multiplexing will allow multiple laser beams to be carried over a single passive optical fibre network, thus further multiplying the capacity of the access network; virtually all business customers will be serviced by a fibre optic access link; and wideband radio distribution within the access network may be used for selective distribution of switched video to home users.

Figure A3.3 shows the penetration of fibre into the access network by around the late 1990s. Note that at this stage, there will still be a strong numerical bias towards the simple copper pair system on the left of the diagram. However, the application of fibre technology will grow rapidly in the latter half of the decade. Apart from the inclusion of a mobile base station in the distribution of the passive optical network, the applications of radio in the access network have been left off this figure to avoid it becoming too complex.

By the end of the decade most services will be delivered through the access network at least partially through fibre. This is illustrated in Figure A3.4. Broadband radio distribution will be in significant use as a part of the fixed access network, and possibly as a delivery vehicle for universal personal telecommunications.

Wavelength division multiplexing will multiply the available fibre capacity, and the possibility of direct distribution of both broadcast and switched video and other wideband services will make fibre to the home a realistic proposition for 'green field' sites. All fixed services will be delivered to business premises over fibre access links.

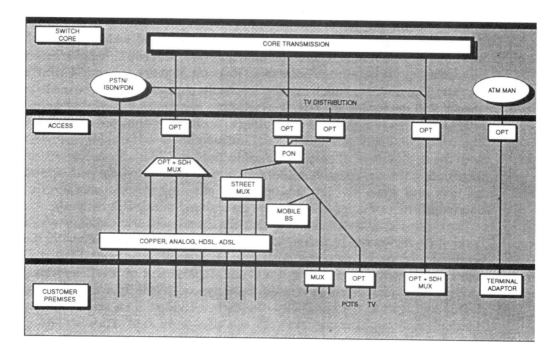

Figure A3.3 Access Network—circa 1995

A3.5 TRANSMISSION—THE HEART OF THE MATTER

The current digital communications network is based on a hierarchy of time division multiplexor equipment, known as the PDH. This system lacks sufficient flexibility to deal with the demands now being placed on the configuration, operation and management of networks, and as a result the telecommunications standards body, the CCITT, has defined a newer and more flexible digital multiplexing scheme, referred to as SDH.

The SDH system takes advantage of more recently developed technology to provide superior inbuilt management and monitoring capabilities, simpler dynamic reconfiguration in response to changing demands or network failures, and higher reliability through a reduction in the number of separate items of equipment typically needed to provide the multiplexing function. SDH also provides a greater maximum link bandwidth (2.5 Gbit/s as against 140 Mbit/s) to economically satisfy the emerging demand for wideband services.

These advantages mean that network operators will be moving to develop their higher-order transmission networks almost entirely using SDH technology, although the older PDH technology will continue to be deployed at lower data rates in the access network for some time.

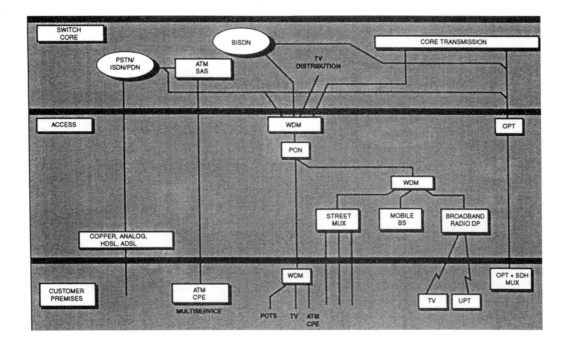

Figure A3.4 Access Network—2000(ish)

With the deployment of SDH, high-order dynamic SDH cross connects will be implemented to cross-connect and multiplex the basic SDH 155 Mbit/s rate into the higher 622 Mbit/s and 2.5 Gbit/s rates. Under the control of the service and network management systems, these will allow a great increase in the flexibility of service provision and restoration.

From the middle of the decade, increasing use will be made of wavelength division multiplexing in the core transmission network to further expand the transmission capacity of fibres. Other advanced optical technologies, such as optical amplification and optical switching, will lead to a reduction in the need for intermediate electronic elements in the fibre transmission network, and allow the implementation of a high-level optical cross-connect switching layer above the SDH layer. The fixed terrestrial transmission core of the network will be totally dominated by optical fibre as the physical transmission medium for the foreseeable future.

A3.6 SWITCHING—THE ROUTE AHEAD

Connecting a user to the service he or she requires has always been one of the main challenges for the designers of wide area networks. Traditionally,

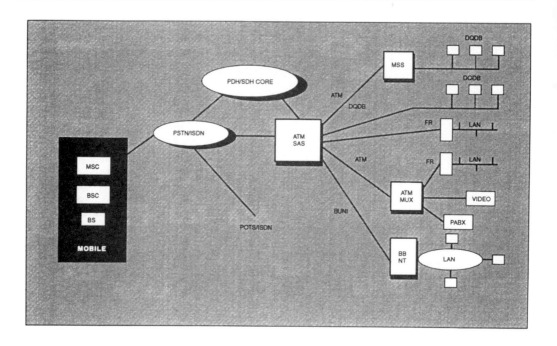

Figure A3.5 Switching Network—circa 1997

switching has been the slowest and/or least flexible part of such networks. The problem has not been shared by local networks; established protocols, such as CSMA/CD and token ring, allow a high-capacity medium to be shared among a large number of users. In this latter case, connecting (for instance) a client to a server has not been a real issue. However, as the facilities and expectations of current LANs grow such that the traditional concerns of the WAN designer take hold, so switching technology will take a more central role.

In the near future, SMDS and ATM based switching systems will be widely deployed to provide wideband services for business customers. The likely situation around the middle of the decade is shown in Figure A3.5.

ATM and SDH based service access switches will begin to be used for large business customers, particularly those requiring wideband services. This will allow all the customer services delivered over the access network to be dynamically reconfigured, and directed to different switching or transmission resources within the network switching and transmission core.

The falling cost of transmission relative to switching and the increasing reliability and redundancy of switching equipment will make the economics of centralising switching nodes irresistible. Over the coming decade, switching equipment will move towards becoming a commodity product.

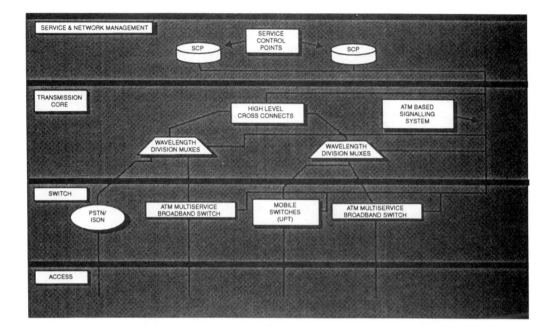

Figure A3.6 Switching Network—2000

The availability of standardised interfaces and sophisticated high-speed signalling systems in the network will enable greater separation of network concerns. For instance, it will be possible for routing decisions and the decision logic required for complex services to be carried out (on a network-wide basis) outside the switching equipment itself. The increasing demand for intelligent network services will promote this separation.

Dynamic cross-connect resources traditionally associated with the transmission network, such as wavelength division multiplexers and high-level cross connects will increasingly be integrated into the network-wide switching strategy, effectively allowing dynamic reconfiguration of the transmission network in response to changing demand on specific routes, or failures within the transmission network.

Towards the end of the decade, the deployment of a true universal broadband switched network based on ATM technology will begin, as shown in Figure A3.6. This will allow the provision of a wide range of switched services to business customers over a single high-speed access link.

A possible first step in deployment of a multiservice network would be the installation of ATM Service Access Switches (SAS) between the switching/transmission networks and the access network.

This will allow rapid provision of a wide range of switched and dedicated services, including wideband services, to large corporate customers over a single highly reliable managed access medium.

Initially the ATM switches could be used to establish PVC connections between CPE equipment and the PSTN/ISDN for switched services, as well as the SDH/PDH transmission core for dedicated services.

In combination with high-level cross-connect switches in the transmission core of the network, and a comprehensive network and service management system, this will allow dynamic reconfiguration of the network to accommodate demand for connection of new services without any physical reconfiguration of the network. This step will also provide a platform for the eventual deployment of B-ISDN to provide switched wideband services. Deployment of this technology will hasten the penetration of ATM-based CPE.

By the end of the decade, the level of complexity in routing and service logic in the switching layer will have been substantially reduced, with routing and service functions being implemented at service control points within the service and network management layer.

Increasing use will be made of cross-connect resources within the transmission network to complement the switching function, and in particular to dynamically reconfigure the transmission network under the control of service and network control points, in response to changes in demand for switched capacity and transmission or switching network failures.

The establishment of a fully functional B-ISDN network based on ATM technology will allow the provision of true wideband multiservice communications, both switched and dedicated within the same integrated network.

Provision of complex intelligent network services and dynamic reconfiguration of network resources will be integrated on a network wide basis under the control of the service and network management systems. The mobile services of the universal personal telecommunications system could also be integrated within this single network umbrella.

The penetration of switching functions into the transmission core through the use of dynamically reconfigurable multiplexers and cross-connects, and into the access network through the use of in-street concentrators, will blur the distinction between the access, switching and transmission layers of the future network.

A3.7   INTELLIGENCE—THE THINKING NETWORK

Network intelligence is additional software provided above basic switch control for path set-up and release to make service functionality independent of the basic switch[1]. The level of network intelligence has increased significantly through the 1990s and will continue to progressively revolutionise the concepts behind service delivery to customers by providing much greater flexibility and service customisation.

---

[1] The seminal thoughts behind intelligent networks were simple—to put the operator role back in place!

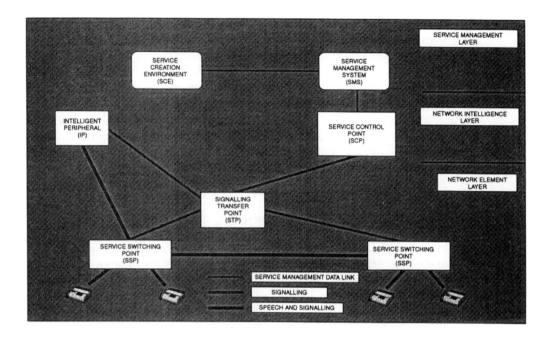

Figure A3.7   Intelligent Network Architecture

The (re)introduction of this capability within the network will progressively reduce the function of the switching elements to the point where probably only the most basic simple calls are handled totally within the switch software. This represents a fundamental reversal of the trends in past years, where switch manufacturers have implemented increasingly complex service logics within the switch software.

The availability of a standardised intelligent network architecture defined by CCITT, standard service definitions, standardised interfaces between switching elements and processing equipment, and standardised computing platform environments will make it possible to construct a sophisticated intelligent network from hardware and software elements obtained from multiple vendors. The importance of standards (both as enablers for the user and as essential features for the vendor) will grow in this area, as they already have in others.

Figure A3.7 illustrates the basis of the intelligent network architecture proposed by CCITT and illustrated in Chapter 6.

Requests for a specific service are transferred from the customer's Service Switching Point (SSP) by the signalling system to the service control point (SCP). This determines the network actions required from the service logic of the required service (e.g. the number translation, time of day routing, etc.) It may also arrange connection of the customer to an intelligent peripheral device (e.g. for digit collection or voice-prompted customer menu selection), or collect information from other remote service control points concerning service logics for the call destination. Once the appropriate actions have been determined, the signalling network is used to pass specific instructions to provide the requested service back to the service switching points.

The service creation environment (SCP) provides tools to the network provider to allow rapid creation of new types of services from reusable software components, and the service management system provides operations support, allowing the updating of SCP information relating to particular customers or particular network services. Implementation of such intelligent network architectures will commence from around the middle of the decade, and will be well advanced by the year 2000.

A3.8 MANAGEMENT—MAINTAINING THE EDGE

Today's networks are characterised by a wide variety of wiring schemes, access methods, protocols, equipment and networking technologies. The range of these grows unabated, and the already wide choice is heading towards an exponential increase in network complexity. Furthermore, there are typically many vendor-specific management systems, with little or no commonality or integration between the various systems.

The ultimate goal in network management evolution will be to achieve a single integrated network management system platform to provide a single point for operational control and allow the automation of all management functions over all network elements. Standard management interfaces, either within the network elements, or in associated element managers or mediation devices will be essential to allow efficient management of practical, disparate networks as a single entity. The increasing use of standard interfaces and platforms also creates the opportunity for reusable software components to be deployed in the network management system. This move towards standardised network management interfaces and operations is already well advanced and is covered in some detail in Chapter 8.

The consolidation of service management techniques is not so far forward. In the absence of an overall defining architecture, various management systems have been developed for each group of services provided. For instance, within most large network operators, there are typically over 100 service management systems operating independently of each other and supporting thousands of products in dozens of identifiable product groups. In addition, some of these service management systems also provide network management functions.

To meet customer expectations for responsiveness and quality of service, an integrated approach to service management which allows effective monitoring and control of the level of customer service will inevitably come to the fore in the development schedules of many suppliers. The way in which network and service management is being integrated is described and illustrated in Chapter 8.

Over the next few years integrated service management systems will be introduced, particularly to assist large business customers in the management of their business communication needs. These systems will then be developed to encompass all business and residential customers.

Throughout the 1990s, service management will increase to become a major component of world-wide telecommunications investment. Knowledge-based systems and artificial intelligence tools will be used, possibly in conjunction with voice recognition and speech synthesis platforms to provide automated service management.

By the year 2000 service management will be well established, with global communications services making use of network components residing in countries around the world to allow customers to control the network to meet their individual telecommunications needs.

Integration of service management systems will need to be facilitated by the use of open systems interfaces and standard information exchange protocols. Future service management systems, along with common network management systems and the implementation of the intelligent network architecture, will facilitate the definition and implementation of new services through the use of computer-aided software engineering (CASE) tools and object oriented programming techniques. These will be used to define new services in a service creation environment, from which information for actual service implementation can be transferred to service control points throughout the network.

A3.9   TWO SNAPSHOTS

It would be wrong to think that all of the above will apply uniformly in the future. There will be many instances in which the level of technology already available is sufficient. To illustrate the range of future networks, here are a couple of 'typical' cases.

Small Inc.

This is a small company that has three sites, all in the same country. Each site has several hundred people who need to communicate frequently and quickly. The LAN facilities at each site provide adequate mail, file transfer and store capability. The flow of information between sites is frequent but not large; nor is it time-critical.

In this case there is little motivation to change; exisiting network technology provides all that is necessary, the network is easy to maintain and cost management is no problem. The network needs of the company are readily catered for with commercially available LANs, routers, modems, leased lines and the like. Basic communications utilities such as Telnet and ftp provide all that is required, and the management facilities provided with the network components allow all the necessary management and administration to be readily carried out in house.

Global Giant Corporation

GGC is at the other end of the spectrum in terms of the demands it places on network technology. Unlike Small Inc., this organisation relies to a large extent on 'information workers'. The competitive edge of the company depends critically on their effectiveness in finding, manipulating and sharing data. The people in GGC are distributed all over the world and many of them either work from home or move frequently from one location to another. The information that passes around the company is usually quite varied and there is a frequent need to import/export data.

Many of the network advances in this appendix are relevant to GGC: they require more bandwidth to their homeworkers, the capacity for exchanging multimedia documentation, network embedded intelligence for rapid reconfiguration, and management systems to keep a disparate installed base operating efficiently. Unlike Small Inc., there are few benefits to be reaped from managing a sophisticated network in-house. It is not the core business of the company, and the cost of building and maintaining a specialist team spread across the exisiting company structure makes little sense. In this case, a virtual private nework places the problem of providing and operating an international network with a specialist who is set up for the job.

The next few years will see a significant change in how GGC operates. First, the new technologies will combine to allow them to operate in a much more flexible way; they will be able to use network services to support the core business, rather than having to bend the business because of network limitations. Future flexibility and capacity will enable mobility, distribution and multimedia communications, all central for GGC's way of working. The second significant change will be that the networks that were provided as a set of independent centres will become a single resource. A combination of rising cost, increasing complexity and diversion from core business will push GGC to outsource network provision and management to a specialist operator. From the information workers' point of view, they will see more of a service to support them than a technology to be grappled with.

A3.10 SUMMARY

Two points are made in this appendix. The first is that previously disparate networks are being overtaken by one multiservice network. The second is that the traditional distinction between private and public networks is fading fast. Both of these will have significant impact on what future networks are capable of and the way that they are provided and run.

To illustrate these two points, the majority of the appendix deals with the technology that the public network operators are developing. The falling cost and proliferation of network implementation and service delivery technologies, together with rising customer expectations and increasing competition/

globalisation are explored as the factors that are driving these network developments.

The five key areas in which new technologies are impacting have been covered. These are as follows.

- The access layer, used to deliver the service to the end customer: this is currently primarily provided by twisted pair copper. The way in which the capacity of the copper will increase has been discussed, along with the likely increase in penetration of fibre in the local network.

- The transmission capacity of the network is provided primarily by digital transmission systems, which are based on standards developed initially for the carriage of digital voice. The implementation of more flexible transmission techniques is already under way, and this has been explained.

- The switch layer is used to provide the various switched services available, which are currently limited to a maximum bit rate of 64 kbit/s. The movement towards broadband switching has been covered as part of the increasing integration of local and wide area network facilities.

- The network management systems currently available tend to be specific to one type of equipment in a very diverse network. The progress towards automated end to end network and service management for disparate networks has been outlined.

- Network intelligence, the additional software provided above basic switch control that allows switch-independent facilities to be implemented, completes the picture of the evolving network of the future.

Together, these technical advances will change the way in which networks are used, managed and provided. Since this impacts on almost everyone in some way, forewarned is forearmed.

REFERENCES

British Computer Society (1989) *The Future Impact of Information Technology.* BCS Trends in IT series.

Gallagher, R. M. (1991) Managing SDH network flexibility. *British Telecom Engineering*, **10**, July.

Monk, P. (1989) *Technological Change in the Information Economy.* Pinter Press.

Public Networks (1994) *Deploying Broadband*, **4**, No. 6, 59–66.

Rudge, A. W. (1994) The infosphere. *BT Engineering Journal*, February.

Spackman, J. W. (1990) The networked organisation. *British Telecom Engineering*, **9**, April.

West, S., Norris, M. and Stockman, S. (1997) *Computer Systems for Global Telecommunications.* Chapman & Hall.

Willets, K. J. (1991) Cooperative management: the key to managing customer networks. *British Telecom Engineering*, **10**, October.

Glossary

A little inaccuracy sometimes saves tons of explanation

H. H. Munro
[Saki]

We list here many of the abbreviations and concepts introduced through the book. Some of the terms that have been explained in the main text are defined below using different words. We hope that this will help clarify some of the more complex ideas—not, as the above quote indicates, perfectly, but at least usefully.

Access control method	A methodology of distinguishing between the different LAN technologies. By regulating each workstation's physical access to the transmission medium, it directs traffic around the network and determines the order in which nodes gain access, so that each user obtains an efficient service. Access methods include Token Ring, Arcnet, FDDI and Carrier Sense Multiple Access with Collision Detection (CSMA/CD), a system employed by Ethernet.
Address	A common term used in both computing and telecommunications for the destination or origination of a connection, for instance a location identifier for an Internet-connected device. Addresses can be logical (131.146.6.11, an IP address for a workstation) or personal (name@organisation.domain, to reach an individual)
Address mask	Also known as a subset mask. It is used to identify which bits in an IP address correspond to the network address and which bits refer to a local terminal.
Address resolution	The conversion of an Internet address into its corre-

sponding physical address (for instance a corresponding Ethernet address).

Aggregate

The total bandwidth of a multiplexed bit stream channel, expressed as bits per second.

Algorithm

A group of defined rules or processes for solving a problem. This might be a mathematical procedure enabling a problem to be solved in a definitive number of steps.

ANSA

Advanced Networked Systems Architecture, a research group established in Cambridge, UK, in 1984 that has had a major influence on the design of distributed processing systems. The ANSA reference manual is the closest thing to an engineer's handbook in this area. Information available via World Wide Web, URL http://ansa.co.uk

API

Application Program Interface—software designed to make a computer's facilities accessible to an application program. All operating systems and network operating systems have APIs. In a networking environment it is essential that various machines' APIs are compatible, otherwise programs would be exclusive to the machines in which they reside.

APPC

Advanced Program to Program Communication—an application program interface developed by IBM. Its original function was in mainframe environments, enabling different programs on different machines to communicate. As the name suggests the two programs talk to each other as equals using APPC as an interface.

APPC/PC

A version of APPC developed by IBM to run a PC-based Token Ring network.

Applet

A mobile application program that can be accessed over a network (typically the Internet). It is self contained, in that it carries its own presentation and processing logic, and can run on whatever type of machine imports it. Although fairly new, applets are being used as plug-in units that form part of a larger application. The concept of the applet is tied to that of Java (a compact and portable interpreted language).

AppleTalk

OSI-compliant protocols that are media independent and able to run on Ethernet, Token Ring and LocalTalk. LocalTalk is Apple Computer's proprietary cabling system for connecting PCs, Macintoshes and peripherals, and it uses the CSMA/CA access method.

Application program	More usually referred to simly as 'application', this is a complete, self-contained program that performs a specific function directly for the user. Editors, spreadsheets and text formatters are common examples of applications. Network applications include clients such as those for FTP, electronic mail and Telnet
Architecture	When applied to computer and communication systems, it denotes the logical structure or organisation of the system and defines its functions, interfaces and procedures.
ARP	Address Resolution Protocol—a networking protocol that provides a method for dynamically binding a high-level IP address to a low-level physical hardware address. This means, for instance, finding a host's Ethernet address from its Internet address. ARP is defined in RFC 826.
Asynchronous	An arrangement where there is no correlation between system time and the data that is exchanged, carried or transmitted over the system. For instance, an asynchronous protocol sends and receives data whenever it wants—there is no link to a master clock. The penalty for this freedom is that extra information has to be added to announce the start and stop of a communication.
Asynchronous data transmission	A data transmission in which receiver and transmitter clocks are not synchronised, each character (word/data block) being preceded by a start bit and terminated by one or more stop bits, which are used at the receiver for synchronisation.
Audit trail	A networking audit trail is a continuous record of a network's activity and is a useful network management tool as it shows how resources are being used and where the problems lie.
B Channel	The ISDN term used to describe the standard 64 kbit/s communications channel.
Bandwidth	The difference between the highest and lowest sinusoidal frequency signals that can be transmitted by a communications channel; it determines the maximum information-carrying capacity of the channel.
Basic rate interface	An ISDN term that describes the two interfaces, 64 kbit/s transmission links and a 16 kbit/s signalling channel,

	referred to as bearer links and the delta channel. Also see ISDN.
Batch processing	In data processing or data communications, an operation where related items are grouped together and transmitted for common processing.
BBS	Bulletin Board System. A computer-based system meeting and announcement system that allows people to both post and view information. Often organised into use groups or centred around a particular subject.
Bits per second	The basic measurement for serial data transmission capacity, abbreviated to bps. Usually has some form of modifier—kbps is thousands of bits per second, Mbps is millions of bits per second. Typically, a domestic user will have an Internet line running at a few tens of kbps. Backbone links are usually 2 Mbps and more.
Bridge	A device or technique used to match circuits, thereby minimising any transmission impairment.
Browser	A program which allows a person to read hypertext information. The browser gives some means of viewing the contents of nodes and of navigating from one node to another. Mosaic, Lynx and Netscape are browsers for the World Wide Web. They act as clients to the array of remote servers on which web pages are hosted.
BSI	British Standards Institution.
CASE	Common Application Service Element—a collection of protocol entities forming part of the application layer, that are used to provide common services.
CCITT (now known as ITU–TS)	Consultative Committee of the International Telegraph and Telephone.
CCTA	Central Computer and Telecommunications Agency. UK-based organisation that deals with a wide range of network and processing activities.
CEN/CENELEC	The two official European bodies responsible for standard setting, subsets of the members of the International Organisation for Standardisation (ISO). The main thrust of their work is functional standards for OSI related technologies.
CEPT	The European Conference of Posts and Telecommunications—an association of European PTTs and network operators from 18 countries. It is the sister organisation

	to CEN/CENELEC. CEPT was originally responsible for the NET's standards which are now under the aegis of standards.
Circuit	An electrical path between two points generally made up by a number of discrete components.
Circuit switching	The method of communications where a continuous path is first established by switching (making connections) and then using this path for the duration of the transmissions. Circuit switching is used in telephone networks and some newer digital data networks.
Client	An object which is participating in an interaction with another object, and is taking the role of requesting (and receiving) the required service.
Client–server	The division of an application into two parts, where one acts as the client (by requesting a service) and the other acts as the server (by providing the service). The rationale behind client–server computing is to exploit the local desktop processing power, leaving the server to govern the centrally held information. This should not be confused with PCs holding their own files on a LAN, as here the client or PC is carrying out its own application tasks.
CMIP/CMIS	Common Management Information Protocol/Service. A standard developed by the OSI to allow systems to be remotely managed.
Connection-oriented	The familiar form of communication on the telephone network. A call is initiated by setting up an end-to-end connection between participants and this connection is kept for the duration of the call. It may not be efficient in terms of network usage, but there are some assurances of delivery.
Connectionless	Refers to a communication where two or more participants do not have a fixed path between them. Each of the packets that constitute the communication looks after its own routing. This arrangement is subject to the vagaries of network availability but can be a very efficient overall way of using a network.
CORBA	Common Object Request Broker Architecture. An evolving framework being developed by the Object Management Group to provide a common approach to systems interworking.
COSE	Common Open Systems Environment.

CPE	Customer Premises Equipment.
CSMA/CD	Carrier Sense Multiple Access with Collision Detection— a method used in local area networks whereby a terminal station wishing to transmit listens and transmits only when the shared line is free. If two or more stations transmit at the same time, each backs off for a random time before retransmission. Each station monitors its signal and if this is different from the one being transmitted, a collision is said to have occurred (collision detection). Each backs off and then tries again later.
D channel	The ISDN term used to describe the standard 16 kbit/s signalling channel. Originally known as the delta channel, but became the D channel as early text processors could not cope with Greek symbols!
Data compression	A method of reducing the amount of data to be transmitted by applying an algorithm to the basic data source. A decompression algorithm expands the data back to its original state at the other end of the link.
Database	A collection of interrelated data stored together with controlled redundancy to support one or more applications. On a network, data files are organised so that users can access a pool of relevant information.
Database server	The machine that controls access to the database using client–server architecture. The server part of the program is responsible for updating records, ensuring that multiple access is available to authorised users, protecting the data and communicating with other servers holding relevant data.
Datagram	A variety of data packet. A self-contained, independent entity of data carrying enough information to be routed from source to destination without reliance on earlier exchanges between the source and destination.
DBMS	DataBase Management Systems—groups of software used to set up and maintain a database that will allow users to call up the records they require. In some cases, DBMS also offer report and application-generating facilities.
DCE	Distributed Computing Environment. A set of definitions and components for distributed computing developed by the Open Software Foundation, an industry-led consortium.
Distributed computing	A move away from large centralised computers such as

minicomputers and mainframes, to bring processing power to the desktop. Often confused with distributed processing.

Distributed database   A database that allows users to gain access to records, as though they were held locally, through a database server on each of the machines holding part of the database. Every database server needs to be able to communicate with all the others as well as being accessible to multiple users.

Distributed processing   The distribution of information processing functions among several different locations in a distributed system.

DNS   Domain Name Service. A general-purpose distributed, replicated, data query service used on the Internet for translating host names into Internet addresses, e.g. taking a dot address such as jungle.pdg.com and returning the corresponding numerical address.

DQDB   Dual queue distributed bus.

ECMA   European Computer Manufacturers Association—an association composed of members from computer manufacturers in Europe; it produces its own standards and contributes to CCITT and ISO.

EDI   Electronic Data Interchange. Basically this refers to agreed formats to enable invoices, bills, orders, etc. to be sent and fulfilled over the network. The term EDI is increasingly being supplanted by the term Electronic Trading.

Electronic mail   Message automatically passed from one computer user to another, often through computer networks and/or via modems over telephone lines. Abbreviated to E-mail.

Electronic mail address   The coding required to ensure that an electronic mail message reaches its specified destination. There are many formats of mail address, perhaps the best known being the dot address used for Internet mail, e.g. 'name@organisation.domain'.

Encapsulation   The transparent enveloping of one protocol within another for the purposes of transport. Encapsulation is, along with tunnelling, a favourite method for supporting multiple protocols across linked networks.

Encryption   A means of turning plain text into cipher text, hence protecting content from eavesdroppers. There are many ways of doing this, public and private key encryption being the two main ones.

Ethernet	A Local Area Network (LAN) characterised by 10 Mbit/s transmission using CSMA/CD—Carrier Sense Multiple Access with Collision Detection.
ETSI	European Telecommunications Standards Institute.
Fast packet switching	A new technology that differs from traditional packet switching. One differing aspect is that it transmits all data in a single packet format, whether the information is video, voice or data.
Fault tolerance	A method of ensuring that a computer system or network is more resilient to faults or breakdowns, to avoid lost data and downtime. Differing applications achieve this, and they include processor duplication and redundant media systems.
FDDI	Fibre Distributed Data Interface—an American National Standards Institute (ANSI) LAN standard. It is intended to carry data between computers at speeds up to 100 Mbit/s via fibre-optic links. It uses a counter-rotating Token Ring topology and is compatible with the first, physical, level of the ISO seven-layer model. FDDI technology boasts a very low bit error rate
FDM	Frequency Division Multiplexing—when a signal is split across frequency bands by modulation.
File server	A station in local area networks dedicated to providing file and data storage to other terminals in the network.
Firewall	In general, refers to the part of a system designed to isolate it from the threat of external interference (both malicious and unintentional).
Firewall machine	A dedicated machine that usually sits between a public network and a private one (e.g. between an organisation's private network and the Internet). The machine has special security precautions loaded onto it and used to filter access to and from outside network connections and dial-in lines. The general idea is to protect the more loosely administered machines hidden behind the firewall from abuse.
FTAM	File Transfer and Manipulation—a protocol entity that forms part of the Application Layer, enabling users to manage and access a distributed file system.
ftp	File transfer protocol. A facility (usually in the form of a software package loaded on a PC) that allows files to be exchanged between remote computers. Widely used to

transfer and access information within the Internet community (see references in Chapter 2).

G.703
The CCITT standard for the physical and logical transmissions over a digital circuit. Specifications include the US 1.544 Mbit/s and the European 2.048 Mbit/s that use the CCITT recommended physical and electrical data interface.

G.703 2.048
Transmission facilities running at 2.048 Mbit/s that use the CCITT recommended physical and electrical data interface.

Gateway
Hardware and software that connect incompatible networks, which enables data to be passed from one network to another. The gateway performs the necessary protocol conversions.

GDMO
Guidelines for the Definition of Managed Objects. Part of the standards set that enable disparate networks to be managed as integrated virtual networks.

Gopher
One of a number of Internet-based services that provide information search and retrieval facilities.

GOSIP
Government Open Systems Interconnect Profiles. UK and US initiatives to help users to procure open systems.

Groupware
A general term to denote software-based tools that can be used to support a distributed set of workers. This covers applications as disparate as Windows for Workgroups through to PC videophones. More formally called Computer Supported Co-operative Working (CSCW).

GUI
Graphical User Interface—an interface that enables the user to select a menu item by using a mouse to point to a graphic icon (small simple pictorial representation of a function such as a paint brush for shading diagrams, etc.). This is an alternative to the more traditional character-based interface where an alphanumeric key-board is used to convey instructions.

Hierarchical network
A network structure composed of layers. An example of this can be found in a telephone network. The lower layer is the local network followed by a trunk (long-distance) network up to the international exchange networks.

Hostage data
Data which are generally useful but are held by a system which makes external access to the data difficult or expensive.

HTML	HyperText Markup Language. HTML is the language used to describe the formatting in World Wide Web documents. As well as text layout, HTML is used to place pictures, insert buttons, and specify links to other documents.
HTTP	HyperText Transfer Protocol. The basic protocol underlying the World Wide Web system. It is a simple, stateless request–response protocol.
Hypertext	A means of presenting documentation so that links to related text are readily apparent. Hypertext systems allow a user to input certain words, pictures or icons and immediately display related information for the selected item. Hypertext requires some form of language (like HTML) to specify branch labels with a hypertext document.
IAB	Internet Activities Board. The influential panel that guides the technical standards adopted over the Internet. Responsible for the widely accepted TCP/IP family of protocols. More recently, the IAB has accepted SNMP as its approved network management protocol.
IEEE	The Institute of Electrical and Electronic Engineers. US-based professional body covering network and computing engineering.
IEE	UK equivalent of the IEEE.
IEEE 802.3	The IEEE's specification for a physical cabling standard for LANs, as well as the method of transmitting data and controlling access to the cable. It uses the CSMA/CD access method on a bus topology LAN, and is operationally similar to Ethernet. Also see OSI.
IEEE 802.4	A later physical standard that uses the token passing access method on a bus topology LAN.
IEEE 802.5	A later physical LAN standard that uses the token passing access method on a ring topology LAN. Used by IBM on its Token Ring systems. Also see OSI.
IETF	Internet Engineering Task Force—a large, open international community of network designers, operators, vendors and researchers whose purpose is to co-ordinate the operation, management and evolution of the Internet and to resolve short- and mid-range protocol and architectural issues. It is a major source of proposals for protocol standards which are submitted to the Internet Architecture Board (IAB) for final approval.

Information processor   A computer-based processor for data storage and/or manipulation services for the end user.

Information retrieval   Any method or procedure that is used for the recovery of information or data which has been stored in an electronic medium.

Information superhighway   A much used (and abused) term that refers to a combination of high speed networks and sophisticated applications for information handling. The term was first coined in the Clinton/Gore administration whose plans to deregulate communication services began with their 1994 legislation to promote the integration of concepts from Internet, telephone providers, business networks, entertainment services, information providers, education, etc.

Intelligent terminal   A terminal that contains a processor and memory with some level of programming facility. The opposite is a dumb terminal.

Internet   The largest network of computers in the world. It actually comprises many smaller networks that use the TCP/IP protocols to communicate and share a common addressing scheme and naming convention. The Internet is growing at a phenomenal rate and has sparked a wealth of technical and social innovation over the years.

Internet address   The 32-bit host address defined by the Internet Protocol (IP) in RFC 791. The Internet address is usually expressed in dot notation, e.g. 128.12.4.5. The address can be split into a network number (or network address) and a host number unique to each host on the network and sometimes also a subnet address. The dramatic growth in the number of Internet users over the last few years has led to a shortage of new addresses. This is one of the issues being addressed by the introduction of a new version of IP, IPv6.

intranet   Usually with a lower case 'i', this term denotes any set of networks interconnected with routers.

IP   The ubiquitous Internet Protocol, one of the key parts of the Internet. IP is a connectionless (i.e. each packet looks after its own delivery), switching protocol. It provides packet routing, fragmentation and reassembly to support the Transmission Control Protocol (TCP). IP is defined in RFC 791.

IP address   The Internet Protocol address. This is a 32-bit address that has to be assigned to any computer that connects to

	the Internet. A typical IP address takes the form 92.61.33.11 and comprises a host component and a network component.
IPv6	The proposed successor to IP. It is a longer address but still compatible with IP. The aims are to extend the available address space, improve security, support user authentication and cater for delay-sensitive traffic.
ISDN	Integrated Services Digital Network—an emerging end-to-end CCITT standard for voice, data and image services. The intention is for ISDN to provide simultaneous handling of digitised voice and data traffic on the same links and the same exchanges.
ISO	International Organisation for Standardisation. ISO is not an abbreviation—it is intended to signify commonality (from Greek *Iso* = same). The ISO is responsible for many data communications standards. A well-known standard produced by ISO is the seven-layer Open Systems Interconnection (OSI) model.
Isochronous	Data transmission in which a transmitter uses a synchronous clock but the receiver does not. The receiver detects messages by start/stop bits, as in asynchronous transmission.
ISP	Internet Service Provider. This is most people's first point of contact with the Internet. An ISP usually offers dial-up access via SLIP or PPP connections to a server on the Internet. Most ISPs also offer their customers a range of client software that can be used on the Net.
IT	Information Technology. A very general term coined in the 1970s to describe the application of computer science and electronics an engineering to the specification, design and construction of information-rich systems.
Java	A compact and portable language that looks as if it will have significant application in the building of highly portable applications (or applets). Java is designed to run on a wide range of computers and to look after its own security and operation. With Java a user can download anything they like the look of over the Internet without the need for all of the software to use it on their local machine.
LAN	Local Area Network—a data communications network

used to interconnect data terminal equipment distributed over a limited area.

LAN Manager | A network operation system developed by Microsoft for PCs attached to Local Area Networks.

Legacy system | A system which has been developed to satisfy a specific requirement and is usually difficult to substantially reconfigure without major re-engineering.

MAN | Metropolitan Area Network. A digital network based on a shared access broadband medium covering an urban area (around 50 km diameter).

MHS | Message handling service—the protocol forming part of the applications layer and providing a generalised facility for exchanging messages between systems.

Mips | Millions of instructions per second—a measure of a computer's processing power is how many instructions per second it can handle.

Modem | MOdulator–DEModulator—data communications equipment that performs necessary signal conversions to and from terminals to permit transmission of source data over telephone and/or data networks.

MPEG | Moving Picture Experts Group. Generally used to refer to coding standards for video images sent over the Internet. MPEG coding is a common standard for which a number of public domain players exist.

Multiplexing | The sharing of common transmission media for the simultaneous transmission of a number of independent information signals; see Frequency Division Multiplexing (FDM) and Time Division Multiplexing (TDM).

Named pipes | Part of Microsoft's LAN Manager—an interface for interprocessing communications and distributed applications. An alternative to NetBios designed to extend the interprocess interfaces of OS/2 across a network.

NetBios | Network Basic input/output system—an IBM protocol. It enables IBM PCs to interface and have access to a network.

NetWare | A Novell LAN operating system and associated products. Novell is a major player in the world LAN server market.

Network | A general term used to describe the interconnection of computers and their peripheral devices by communications channels, for example the Public Switched Telephone Network (PSTN), Packet Switched Data Network (PSDN), Local Area Network (LAN), Wide Area Network (WAN).

Network interface
The circuitry that connects a node to the network, usually in the form of a card fitted into one of the expansion slots in the back of the machine; it works with the network software and operating system to transmit and receive messages on the network.

Network management
A general term embracing all the functions and processes involved in managing a network, including configuration, fault diagnosis and correction. It also concerns itself with statistics gathering in network usage.

Network topology
The geometry of the network relating to the way the nodes are interconnected.

NNTP
Network News Transfer Protocol. A protocol defined in RFC 977 for the distribution, enquiry, retrieval and posting of Usenet news articles over the Internet. It is designed to be used between a newsreader client and a news server.

Non-proprietary
Software and hardware that are not bound to one manufacturer's platform. Equipment that is designed to specification that can accommodate other companies' products. The advantage of non-proprietary equipment is that a user has more freedom of choice and a larger scope. The disadvantage is that when it does not work, you may be on your own.

NT
Network termination. A piece of equipment (usually on a user's premises) that provides network access via a standard interface.

Object
An abstract encapsulated entity which provides a well-defined service via a well-defined interface.

Object orientation
An increasingly popular approach to the design of netwrok systems in which they are composed of a set of objects. Each object is an independent element with defined interfaces and actions. There is a significant formal basis behind this simple idea.

ODP
Open Distributed Processing.

OMG
Object Management Group.

OMNI
Open Management Interoperability. An ISO-based

network management standards body. Responsible for OMNIPoint, which includes CMIS and CMIP.

Open system
A general term for systems that are built with standard interfaces which allow components from different manufacturers to be connected together.

Operating system
Software such as DOS, OS/2, NetWare, VMS that manages the computer's hardware and software. Unless it intentionally hands over to another program, an operating system runs programs and controls peripherals.

OSCA
Open Systems Cabling Architecture—structured cabling system, primarily for local networks, that converts a bus layout to a hub layout.

OSI
Open Systems Interconnection—the ISO Reference Model consisting of seven protocol layers. These are the application, presentation, session, transport, network, link and physical layers. The concept of the protocols is to provide manufacturers and suppliers of communications equipment with a standard that will provide reliable communications across a broad range of equipment types.

Packet
A unit of data sent across a network. A packet usually has a payload—the information it carries—and an overhead, the latter being the extra bits used to get it safely across the network to its destination.

Packet switching
The mode of operation in a data communications network whereby messages to be transmitted are first transformed into a number of smaller self-contained message units known as packets. Packets are stored at intermediate network nodes (packet-switched exchanges) and are reassembled into a complete message at the destination. A CCITT recommendation standard for packet switching is X.25.

PAD
Packet Assembler and Disassembler—a device used in the X.25 packet switched network, permitting terminals which cannot interface with the network to do so. PAD converts terminals data to/from packets and handles call set-up and addressing.

PCMCIA
Personal Computer Memory Card International Association. This organisation was set up in 1989 to establish standards for personal computer cards—small, plug-in units designed to be installed or removed without

opening the computer case which provide additional memory and input/output functions. Three types of card are currently specified. Type 1 is 3.3 mm thick and is designed to provide extra memory. Type 2 is 5 mm thick and is used, typically, for modems. Type 3 at 10.5 mm is intended for special needs such as high capacity memory.

PDH — Plesiochronous Digital Hierarchy. This usually refers to the layers of a public operator's transmission network.

Peer to peer — Communications between two devices on an equal footing, as opposed to host/terminal or master/slave. In peer to peer communications both machines have and use processing power.

Pipe — Installed in most operating systems, a pipe is a method used by programs to communicate with each other. One of OS/2's attributes is that pipes can be created quickly and easily. When a program sends data to a pipe, it is transmitted directly to the other program without ever being written onto a file.

Pixel — Picture element—the smallest discrete element making up a visual display image.

Point-to-point — Direct link between two points in a network or communications link.

Polling — Process of interrogating terminals in a multipoint network in turn in a prearranged sequence by controlling the computer to determine whether the terminals are ready to transmit or receive. If so, the polling sequence is temporarily interrupted while the terminal transmits or receives.

PON — Passive Optical Network.

PoP — Point of Presence. A site where there exists a collection of telecommunications equipment, usually modems, digital leased lines and multi-protocol routers. The PoP is put in place by an Internet Service Provider (ISp). An ISP may operate several PoPs distributed throughout their area of operation to increase the chance that their subscribers will be able to reach one with a low cost telecommuniations circuit. The alternative is for them to use virtual PoPs (virtual points of presence) via some third party.

Port — A device which acts as an input/output connection. Serial port or parallel port are examples.

POSIX	Portable Operating System Interface to Unix. One of the standardisation initiatives in the distributed processing area. The aim is to allow software to be moved between proprietary hardware with minimal need for rework.
Proprietary	Any item of technology that is designed to work with only one manufacturer's equipment. The opposite of the principle behind Open Systems Interconnection (OSI).
Protocol	A set of rules and procedures that are used to formulate standards for information transfer between devices.
Protocol relay	A relaying function which operates by mapping the elements of the protocol on one side of the relay system to those on the other.
PSTN	Public Switched Telephone Network—the public telephone system providing local, long-distance and international telephone services. In addition, widely used (with modems) for many other data services.
PTT	Postal, Telegraph and Telephone—the administrative authority in a country that controls all postal and public telecommunication services in that country. It is the same as PNO—Public Network Operator.
PVC	Private Virtual Circuit.
Quality of service	Measure of the perceived quality of a service. Usually based on tangible metrics such as time to fix a fault, average delay, loss percentages, system reliability, etc.
RARP	Reverse Address Resolution Protocol. A protocol which provides the reverse function of ARP. RARP maps a hardware address to an Internet address.
RFC	Request For Comment. The main vehicle for the publication of Internet standards, such as SNMP.
RM-ODP	Reference Model for Open Distributed Processing.
ROSE	Remote Operations Service Element—a protocol code forming part of the applications layer, providing facility for initiating and controlling operations remotely.
Router	A router operates at level 3 of the OSI model. Routers are protocol specific and act on routing information carried out by the communications protocol in the network later. A router is able to use the information it has obtained about the network topology and can

choose the best route for packet to follow. Routers are independent of the physical level (layer 1) and can be used to link a number of different network types together.

Routing   The selection of a communications path for the transmission of information from source to destination.

RPC   Remote Procedure Call. A protocol which allows a program running on one host to cause code to be executed on another host without the programmer needing to explicitly code for this. RPC is a popular option for implementing the client–server model of distributed computing.

RS-232   The most common serial line standard. It uses 25-way D-type connectors but often only three wires are connected—one to ground (pin 7) and one for data in each direction.

SASE   Specific Application Service Element—a collection of protocol codes forming part of the application layer for specific services such as file and job transfers.

Screen scraping A method of accessing a server where the client presents itself as being a direct interface to a human user. The client reads information from the screen presented by the server and sends information as keystrokes from the pretend user.

SDH   Synchronous Digital Hierarchy. This is tending to supersede PDH as the core transmission approach used by public operators. SDH provides greater control and flexibility over the structure of digital links by including extra information in the transmission frame structure.

Server   An object which is participating in an interaction with another object (usually a client), and is taking the role of providing the required service. One half of client—server system.

Service   An independently useful and well-defined function.

Session   The connection of two nodes on a network for the exchange of data—any live link between any two data devices.

SGML   Standard Graphical Markup Language. An international standard encoding scheme for linked textual information. HTML is a subset.

Signalling   The passing of information and instructions from one point to another for the setting up or supervision of a telephone call or message transmission.

SMTP

Simple Mail Transfer Protocol. The Internet standard for the transfer of mail messages from one processor to another. The protocol details the format and control of messages.

SNA

Systems Network Architecture—an IBM layered communications protocol to send data between IBM hardware and software.

SNMP

Simple Network Management Protocol—consists of three parts: Structure of Management Information (SMI), Management Information Base (MIB) and the protocol itself. The SMI and MIB define and store the set of managed entities; SNMP transports information to and from these entities.

Socket

A mechanism for creating a virtual connection between processes. At the simplest level, an application opens the socket, specifies the required service, binds the socket to its destination and then sends or receives data.

Software

All programs (plus documentation) which are associated with a computer or computer-based system, as opposed to hardware which is the physical equipment.

Sonet

A synchronous optical transmission protocol. Sonet is intended to be able to add and drop lower bit rate signals from the higher bit rate signal without needing demultiplexing. The standard defines a set of transmission rates, signals and interfaces for fibre-optic transmission.

SQL

Structured query language. A widely used database access language.

Store and forward

A technique used in data communications in which message packets are stored at an intermediate node in a network and then forwarded to the next routing point, where an appropriate line becomes free.

Switched network

A network which is shared by several users, any of whom can establish communication with any other by means of suitable interconnection (switching) operations.

Switching

Process by which transmissions between terminals are interconnected, effected at exchange at nodal points in the network.

Synchronisation

The actions of maintaining the correct timing sequences

for the operation of a system.

Synchronous transmission	Transmission between terminals where data are normally transmitted in blocks of binary digit streams, and transmitter and receiver clocks are maintained in synchrony. User-controlled process by which remote terminals are temporarily interconnected for the transfer of information. It embraces store- and forward-switching as used in packet-switched and cell-switched networks and circuit-switching.
System	A collection of independently useful objects which happen to have been developed at the same time.
TA	Terminal Adaptor. A piece of equipment used with an ISDN connection to allow existing terminals to hook up. The equivalent of a modem.
TCP	Transmission Control Protocol. The most common transport layer protocol used on Ethernet and the Internet. It was developed by DARPA. TCP is built on top of Internet Protocol (IP) and the two are nearly always seen in combination as TCP/IP (which implies TCP running on top of IP). The TCP element adds reliable communication, flow control, multiplexing and connection-oriented communication to the basic IP transport. It is defined in RFC 793.
TCP/IP	Transmission Control Protocol/Internet Protocol. The set of data communication standards adopted, initially on the Internet, for interconnection of dissimilar networks and computing systems.
TDM	Time Division Multiplexing—when a signal is split across time slots in a composite transmission channel.
Telecommunications	The general name given to the means of communicating information over a distance by electrical and electromagnetic methods. The transmission and reception of information by any kind of electromagnetic system.
Teleworking	Using computing and communications technology to work away from an office.
Telnet	A TCP/IP-based application that allows connection to a remote computer.
Throughput	A way of measuring the speed at which a system, computer or link can accept, handle and output information.

TINA	Telecommunications Intelligent Network Application. A consortium that aims to ease the integration of telecommunications and computing technology and promote the development of intelligent networks.
Topology	A description of the shape of a network; for example, star, bus and ring. It can also be a template or pattern for the possible logical connections onto a network.
TP	Transaction processing. Concerned with controlling the rate of enquiries to a database. Specialist software (known as a TP monitor) that allows a potential bottleneck to be managed.
Transmission	The act of transmitting a signal by electrical/electromagnetic means over a communications channel.
Tunnelling	Usually refers to a situation where a public network is used to connect two private domains so that the privacy of the overall link is maintained. To all intents and purposes, the same as encapsulation.
UPT	Universal Personal Telephony.
URL	Uniform Resource Locator. A standard for locating an object on the Internet and most widely known as the form of address for pages on the World Wide Web. Typical URLs take the form http://www.identity.com/, ftp://archive.ic.ac/fred or telnet://jungle.com. The part before the first colon specifies the access scheme or protocol. The part after the colon is interpreted according to the access scheme. In general, two slashes after the colon indicate a host name.
Usenet	Probably the largest decentralised information utility in the world. It encompasses government agencies, universities, schools, businesses and hobbyists. Hosts well oer 10 000 special interest groups and incorporates the equivalent of several thousand paper pages of new technical articles, news, discussion, opinion, etc., every day.
UUCP	Unix–Unix Communication Protocol. A basic mechanism that allows computers running the Unix operating system to interoperate.
Vendor independent	Hardware or software that will work with hardware and software manufactured by different vendors—the opposite of proprietary.
Virtual circuit	A transmission path through a packet-switched data

	network (PSDN), established by exchange of set-up messages between source and destination terminals.
Virtual device	A module allocated to an application by an operating system or network operating system, instead of a real or physical one. The user can then use a computer facility (keyboard or memory or disk or port) as though it were really present. In fact, only the operating system has access to the real device.
VPN	Virtual Private Network. A collection of switching, transmission and customer premises equipment (usually of different types and from different suppliers) that combine to provide a coherent network. From the user's point of view a VPN is a single dedicated network —a service. From the owner's viewpoint, the VPN is a resource that can be managed as an integrated whole.
Window	A flow control mechanism the size of which determines the number of data units that can be sent before an acknowledgement of receipt is needed, and before more can be transmitted.
Windows	A way of displaying information on a screen so that users can do the equivalent of looking at several pieces of paper at once. Each window can be manipulated for closer examination or amendment. This technique allows the user to look at two files at once or even to run more than one program simultaneously.
World Wide Web	Also referred to as the Web, WWW and W3. It is the Internet-based distributed information retrieval system that uses hypertext to link multimedia documents. This makes the relationship of information that is common between documents easily accessible and completely independent of phyiscal location. WWW is a cli-ent–server. The client software takes the form of a 'browser' that allows the user to easily navigate the information on-line. Well-known browsers are Netscape and Mosaic. A huge amount of information can be found on World Wide Web servers.
X.25	A widely used standard protocol suite for packet switching that is also approved by the ISO. X.25 is the dominant public data communications packet standard. Many X.25 products are available on the market and networks have been installed all over the world.
X.400	A store and forward Message Handling System (MHS)

standard that allows for the electronic exchange of text as well as other electronic data such as graphics and fax. It enables suppliers to interwork between different electronic mail systems. X.400 has several protocols, defined to allow the reliable transfer of information between User Agents and Message Transfer Agents.

X.500   A directory services standard that permits applications such as electronic mail to access information, which can be either central or distributed.

X/Open   An industry standards consortium that develops detailed system specifications, drawing on available standards.

Index